Energy Harvesting with Functional Materials and Microsystems

Devices, Circuits, and Systems

Series Editor
Krzysztof Iniewski
CMOS Emerging Technologies Research Inc.,
Vancouver, British Columbia, Canada

FORTHCOMING TITLES:

MIMO Power Line Communications: Narrow and Broadband Standards, EMC, and Advanced Processing
Lars Torsten Berger, Andreas Schwager, Pascal Pagani, and Daniel Schneider

Mobile Point-of-Care Monitors and Diagnostic Device Design
Walter Karlen and Krzysztof Iniewski

Nanoelectronics: Devices, Circuits, and Systems
Nikos Konofaos

Nanomaterials: A Guide to Fabrication and Applications
Gordon Harling and Krzysztof Iniewski

Nanopatterning and Nanoscale Devices for Biological Applications
Krzysztof Iniewski and Seila Selimovic

Nanoscale Semiconductor Memories: Technology and Applications
Santosh K. Kurinec and Krzysztof Iniewski

Radio Frequency Integrated Circuit Design
Sebastian Magierowski

Semiconductor Device Technology: Silicon and Materials
Tomasz Brozek and Krzysztof Iniewski

Smart Grids: Design, Strategies, and Processes
David Bakken and Krzysztof Iniewski

Soft Errors: From Particles to Circuits
Jean-Luc Autran and Daniela Munteanu

Technologies for Smart Sensors and Sensor Fusion
Kevin Yallup and Krzysztof Iniewski

VLSI: Circuits for Emerging Applications
Tomasz Wojcicki and Krzysztof Iniewski

Energy Harvesting with Functional Materials and Microsystems

Edited by
Madhu Bhaskaran • Sharath Sriram
Krzysztof Iniewski

CRC Press
Taylor & Francis Group
Boca Raton   London   New York

CRC Press is an imprint of the
Taylor & Francis Group, an **informa** business

CRC Press
Taylor & Francis Group
6000 Broken Sound Parkway NW, Suite 300
Boca Raton, FL 33487-2742

First issued in paperback 2017

© 2014 by Taylor & Francis Group, LLC
CRC Press is an imprint of Taylor & Francis Group, an Informa business

No claim to original U.S. Government works

Version Date: 20130715

ISBN 13: 978-1-4665-8723-6 (hbk)
ISBN 13: 978-1-138-07410-1 (pbk)

Library of Congress Cataloging-in-Publication Data

Energy harvesting with functional materials and microsystems / editors, Madhu
   Bhaskaran, Sharath Sriram, and Krzysztof Iniewski.
      pages cm -- (Devices, circuits, and systems)
   Includes bibliographical references and index.
   ISBN 978-1-4665-8723-6 (alk. paper)
   1. Energy harvesting--Materials. 2. Power electronics. 3. Microelectronics--Power
supply. I. Bhaskaran, Madhu, editor of compilation. II. Sriram, Sharath, editor of
compilation. III. Iniewski, Krzysztof, 1960- editor of compilation.

   TK2896.E74 2013
   621.31'24--dc23   2013022798

Visit the Taylor & Francis Web site at
http://www.taylorandfrancis.com

and the CRC Press Web site at
http://www.crcpress.com

Contents

Preface

Energy harvesting is the capture of external energy from various naturally occurring or prevalent sources. This harnessed energy is stored for or directly used in numerous applications, most particularly electronics and electrical systems. The scope, significance, and impact of energy harvesting research have generated a lot of interest within the scientific community and the general populace at large. A number of major advances in energy harvesting have relied on the identification and application of functionalities of existing materials or microsystems in innovative manners or with adapted designs.

This book aims to provide an overview of recent research advances utilizing functional materials and microsystems for energy harvesting. The emphasis of the coverage is on materials and devices, with resulting technologies capable of powering ultralow-power implantable sensors up to large-scale electrical grids. The book is presented as a collection of carefully selected contributions from international researchers from both academia and industry.

The nine contributed chapters cover four key themes, starting with design of microsystems, which enable and manage energy harvesting, before moving on to chapters dealing with functional materials that underpin major advances in energy harvesting technologies for thermal, solar, and mechanical energy.

Chapters 1 and 2 present approaches and circuit designs for energy harvesting, storage, and use. Chapter 3 reviews opportunities for the combination of thermoelectric materials and microsystems to contribute to energy harvesting, while Chapter 4 covers a breakthrough thermoelectric material technology utilizing thermopower waves as energy sources. Chapters 5 through 7 present a variety of solar cell configurations spanning polymer and monocrystalline silicon solar cells as energy sources for low-power electronics and with approaches for optimization of energy capture and conversion. Chapters 8 and 9 deal with harnessing mechanical energy from ambient sources using piezoelectric vibrational energy harvesters. Thin film materials for piezoelectric energy harvesting and a review of the designs and limitations of this approach are covered.

We hope this coverage of functional materials and microsystems for energy harvesting provides readers with an appreciation of the new and emerging research areas, as well as inspires further new ideas in a very significant research field.

Madha Bhaskaran and Sharath Sriram
RMIT University

Krzysztof Iniewski
CMCS Emerging Technologies Research, Inc.

The Editors

Dr. Madhu Bhaskaran received her PhD from RMIT University, Melbourne, Australia in 2009, following B.E. and M.Eng. degrees in 2004 and 2005, respectively. She is currently a senior research fellow and joint leader of the Functional Materials and Microsystems Research Group at RMIT University. She is an expert on silicon-based fabrication techniques and thin film patterning for the development of microdevices. This is in addition to her expertise in thin film characterization using a variety of spectroscopy, microscopy, and diffraction techniques. This combination of microfabrication and materials science expertise has enabled her to demonstrate novel applications of functional materials.

She has published 90 peer-reviewed publications, including about 55 journal articles in the last 5 years. She is a recipient of the highly competitive Australian Research Council post-doctoral fellowship 2010–2013 for research on energy harvesting from piezoelectric thin films. In 2011, she received worldwide media coverage, including national television and international radio, for her use of *in situ* nanoindentation in characterizing the nanoscale piezoelectric energy generation properties of thin films. She can be reached at madhu.bhaskaran@gmail.com.

Dr. Sharath Sriram is a senior research fellow and joint leader of the Functional Materials and Microsystems Research Group at RMIT University. His expertise includes the synthesis and characterization of functional thin films, underpinned by skills in microelectronic fabrication techniques. He received his PhD from RMIT University in 2009 for work on synthesis and characterization of high-performance and CMOS-compatible piezoelectric thin films. This included the development of the first techniques for quantitative determination of piezoelectric response coefficients of thin films utilizing *in situ* nanoindentation and atomic force microscopy. These research outcomes were recognized by the 2010 gold medal for excellence in research by the Australian Institute of Nuclear Science and Engineering and the 2012 NMI prize from the National Measurement Institute. He is also a recipient of a 2012 Victoria fellowship.

Dr. Sriram has published over 90 peer-reviewed publications, including articles in leading nanoscience journals, including *Journal of the American Chemical Society, ACS Nano, Advanced Functional Materials,* and *Nanoscale,* and has received over A$1.4 million in research and infrastructure funding over the last 3 years. He is a current recipient of the Australian Research Council post-doctoral fellowship for the period 2011–2014, with a research focus on microdevices for chemical sensing and dynamic manipulation of surface nanostructures utilizing *in situ* techniques with electrical stimuli. He can be reached at sharath.sriram@gmail.com.

Dr. Krzysztof (Kris) Iniewski is managing R&D at Redlen Technologies Inc., a start-up company in Vancouver, Canada. Redlen's revolutionary production process for advanced semiconductor materials enables a new generation of more accurate, all-digital, radiation-based imaging solutions. Kris is also a president of CMOS Emerging Technologies Research (www.cmosetr.com), an organization of high-tech events covering communications, microsystems, optoelectronics, and sensors.

In his career, Dr. Iniewski held numerous faculty and management positions at the University of Toronto, University of Alberta, SFU, and PMC-Sierra Inc. He has published over 100 research papers in international journals and conferences. He holds 18 international patents granted in the United States, Canada, France, Germany, and Japan. He is frequently an invited speaker and has consulted for multiple organizations internationally. He has written and edited several books for IEEE Press, Wiley, CRC Press, McGraw–Hill, Artech House, and Springer. His personal goal is to contribute to healthy living and sustainability through innovative engineering solutions. In his leisure time, Kris can be found hiking, sailing, skiing, or biking in beautiful British Columbia. He can be reached at kris.iniewski@gmail.com.

The Contributors

Jeydmer Aristizabal
Simon Fraser University
Burnaby, Canada

Yindar Chuo
Simon Fraser University, Burnaby,
    Canada
Nanotech Security Corp., Vancouver,
    Canada

Jordi Colomer-Farrarons
University of Barcelona
Barcelona, Spain

A. Dompierre
Université de Sherbrooke
Sherbrooke, Quebec, Canada

L. G. Fréchette
Université de Sherbrooke
Sherbrooke, Quebec, Canada

Sasan V. Grayli
Simon Fraser University
Burnaby, Canada

Siamack V. Grayli
Simon Fraser University
Burnaby, Canada

Terry J. Hendricks
Battelle Memorial Institute
Columbus, Ohio, and Corvallis, Oregon

Joanne Huang
Synopsys, Inc.
Mountain View, California

E. Juanola-Feliu
University of Barcelona
Barcelona, Spain

Kourosh Kalantar-Zadeh
RMIT University
Melbourne, Australia

Bozena Kaminska
Simon Fraser University, Burnaby,
    Canada
Nanotech Security Corp., Vancouver,
    Canada

Isaku Kanno
Kobe University
Kobe, Japan

Clint Landrock
Simon Fraser University, Burnaby,
    Canada
Nanotech Security Corp., Vancouver,
    Canada

Vivien Lo
Nanotech Security Corp.
Vancouver, Canada

Purna P. Maharjan
South Dakota State University
Brookings, South Dakota

Pere Ll. Miribel-Català
University of Barcelona
Barcelona, Spain

Victor Moroz
Synopsys, Inc.
Mountain View, California

Badr Omrane
Simon Fraser University, Burnaby,
    Canada
Nanotech Security Corp., Vancouver,
    Canada

Qiquan Qiao
South Dakota State University
Brookings, South Dakota

Gabriel Alfonso Rincón-Mora
Georgia Institute of Technology
Atlanta, Georgia

J. Samitier
University of Barcelona, Barcelona,
    Spain
Institute for Bioengineering of
    Catalonia, Barcelona, Spain
CIBER-BBN, Zaragoza, Spain

S. Vengallatore
McGill University
Montreal, Quebec, Canada

Sumeet Walia
RMIT University
Melbourne, Australia

1 Powering Microsystems with Ambient Energy

Gabriel Alfonso Rincón-Mora

CONTENTS

Abstract: The demand for portable, lightweight, and long-lasting electronics is relentless, filling a growing need in the military to lighten and extend reconnaissance mission work; in space exploration for remote sensors; in biomedical applications for monitoring, prognosis, treatment, and others; and in consumer electronics for disposable and rechargeable everyday products. Conforming to microscale dimensions means that energy and power supplies, conditioning and processing microelectronics, sensors, wireless transceivers, and other constituent subsystems must synergistically share a common miniaturized platform. Integrating and managing microsources, however, present a myriad of diverse and interdependent mechanical, chemical, and electrical challenges. One pivotal constraint is that ultrasmall systems cannot store the energy required to sustain practical lifetimes, which is why energy harvesters have gained so much attention. This chapter aims to illustrate how to energize and power microsystems from ambient energy by reviewing system requirements, describing the state of the art in miniaturized energy-harnessing transducers, and presenting energy-harvesting supply and charger microelectronics currently under investigation and development.

1.1 MICROSYSTEMS

Considering today's growing demand for and operational needs of portable microelectronics is critical in drafting a strategy for supplying a derivative application space that is barely emerging, like wireless microsensors. The challenge is to understand today's market and needs well enough to predict the challenges society will face at the time that research reaches a commercialization stage. Equally important is comprehending not only the state-of-the-art of technologies needed to build such systems, but also the trends that drive them so that today's efforts can have a chance at leveraging the future benefits of technologies currently under development. Ultimately, though, predictions of this sort, however good they may be, amount to speculation that scientists and engineers must continually fine-tune over time and with experience.

1.1.1 Market Demand

Consumers continue to enjoy and, in consequence, demand smaller and more functionally dense products. Unfortunately, tiny devices necessarily constrain sources to small spaces, limiting the amount of energy they store and how much power they supply to impractically low levels. Consider, to cite an illustrative example, that 100 μW, which is drastically below what wireless telemetry requires today to transmit over a distance of 100 m, completely discharges a $5 \times 3.8 \times 0.037$-$cm^3$ 1 mA·Hr thin-film lithium ion battery (Li ion) in 10 h. Adding functionality, as in the case of cellular phones and the applications they now support, only exacerbates the issue because the extra power overhead drains a source that is already easily exhaustible even faster. Although today's efforts aim at reducing the energy that each additional function requires, the cumulative load that wireless transmission and several low-power functions present to a miniature source is nonetheless substantial, reducing the operational life of the system considerably, to the point at which consumers lose interest.

Perhaps a more significant application space emerges in the wake of integration in the form of *in situ* but nonintrusive devices, such as biomedical implants. While the motivation for and benefits of monitoring biological activity with *in vivo* contraptions without frequently replacing or recharging batteries are for the most part obvious, the commercial and societal advantages of retrofitting low-overhead, noninvasive intelligence into expensive and difficult-to-replace technologies like the power grid are arguably more economically, sociologically, and ecologically profound. Feeding performance metrics and the use of energy in a factory into a central processor, for instance, to determine the optimal use of power cannot only save energy on a grander scale when applied across a wide region, but also reduce the emission of environmentally harmful by-products and the need for politically charged petroleum. From a commercial standpoint, the potential ubiquity for such miniature sensors across factories, hospitals, airports, farms, homes, subway stations, shopping malls, and other such centers rivals and quite likely exceeds that of cellular phones at maybe 10 to 100 sensors for every 10×10 m^2 of surface area.

1.1.2 ENERGY AND POWER REQUIREMENTS

Modern microelectronic systems necessarily embed both analog and digital functions—analog because they must not only manage and process continuous real-life signals (e.g., seismic activity, sound waves, motion, temperature, pressure, and others) but also draw energy and condition power from nonideal sources. Stateof-the-art applications, however, also include digital blocks because binary processes enjoy considerably higher noise margins, which is another way of saying they exhibit higher signal-to-noise ratios (SNRs). The fact is that a small voltage variation, say, from 20 to 400 mV, across a 0–1.8 V digital signal has little to no impact on the output word and its bit-error rate (BER), whereas the same variation in a rail-to-rail analog signal swing essentially eradicates more than 1% to 22% of the data. All this is to say the system draws quiescent and switching power from the source, as Figure 1.1 shows [1].

Because the going trend is to incorporate more functionality into a product, a fixed supply voltage is no longer optimal, especially when seeking to extend operational life. While digital-signal processors (DSPs), for example, may draw milliwatts at 1 V, voltage-controlled oscillators (VCOs) and analog–digital converters (ADCs) can demand considerably less power at maybe 1.8 V and power amplifiers (PAs) significantly more power at 3 V. The end result is a system that, to operate optimally, includes multiple supply voltages capable of feeding diverse power levels.

In the end, irrespective of the supply voltages needed, featuring multiple tasks ultimately burdens a source with additional loads. The problem with a miniature source is that energy is scarce and power levels are low. One way of reducing power in the system is to duty-cycle or time-division multiplex tasks across time so that no more

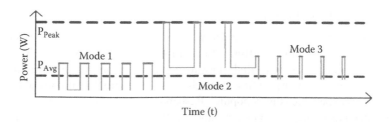

FIGURE 1.1 General loading profile that a modern microsystem presents to a source over time.

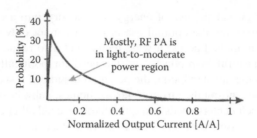

FIGURE 1.2 Probability-density curve for radio-frequency (RF) power amplifiers (PAs) in CDMA (code division multiple access) handset applications.

than one set of interdependent functions operates at any given point. Similarly, powering blocks only when absolutely necessary, as in on-demand, reduces energy that the system would otherwise waste. As a result, systems today dynamically change operating modes *on the fly,* as Figure 1.1 illustrates, where all average and peak power, pulse width, and switching frequency change according to the tasks performed.

Extending the operational life of smart microsensors remains a challenge, even after power-moding and duty-cycling functions. The point is that marketable *microsystems* must draw little to no power. Luckily, portable and sensor applications naturally call for low duty-cycle operation because devices need not function continuously. The cellular phone, for instance, is for the most part alert—that is, ready to receive calls—and while transmission demands milliwatts, awareness only requires microwatts, which is why radio-frequency (RF) power amplifiers are mostly in the light-to-moderate power region of operation, as Figure 1.2 demonstrates [2]. Said differently, while talking continuously on the phone for 3 to 5 hours exhausts a fully charged battery, idling might do so in 5 to 7 days.

While energy and power unavoidably relate, meeting the energy requirements of a microsystem does not necessarily imply the source can also supply the power needed. This means that reducing losses to save energy, for example, by shutting off unused circuit blocks, is as important as duty-cycling tasks over time to decrease power. The source and the circuits that manage the source must therefore account for and respond to various mixed-signal modes whose average and peak power levels, pulse widths, rise and fall times, and switching frequency change over time.

1.1.3 TECHNOLOGY TRENDS

Trends in technology dictate how and to what extent emerging applications succeed. Not surprisingly, the public's affinity to small microelectronic gadgets is motivating scientists and engineers to find innovative ways of confining products into tiny spaces. Accordingly, incorporating as much as possible into the silicon die has been and continues to be increasingly popular. Research and industry are not only building more semiconductor devices on a common substrate, but also postprocessing copper layers, microelectromechanical systems (MEMS), and MEMS-like inductors on top.

The problem with system-on-chip (SoC) integration is higher cost because arbitrarily adding processing steps of increasing complexity to the fabrication process diminishes the commercial appeal of the product. As a result, engineers are also copackaging technologies, for example, like high-frequency gallium–arsenide (GaAs) dies with low-cost complementary metal–oxide semiconductor (CMOS) dies. Power integrated circuit (IC) designers are similarly integrating relatively mature and, therefore, lower cost discrete 1–10 µH power inductors in the $2 \times 2 \times 1$ mm^2 range with their controlling CMOS and bipolar circuits. SoC and system-in-package (SiP) integration, however, are often insufficient, so engineers are also exploring system-on-package (SoP) strategies, for example, by attaching antennae and thin-film Li ions to the external surface of the material encapsulating the silicon die.

Higher SoC integration means machinery with finer photolithographic resolutions, for example, of 45 nm semiconductor devices with thinner barriers and smaller junctions of increasing doping concentration. The resulting electric fields are more intense and their corresponding breakdown voltages are consequently lower, on the order of 1–1.8 V, constraining circuits to operate with lower supply voltages. While reducing the voltages across charging and discharging parasitic capacitors and steady-state loads mitigates power losses, a low supply also decreases usable dynamic range, which translates to a lower SNR and, more generally, reduced noise margin.

Unfortunately, while additional on-chip functions inject and cross couple more noise, SoC, SiP, and SoP designs suffer from diminished filtering capabilities, challenging engineers to design with even lower dynamic ranges. The problem is that, unlike in discrete printed circuit-board (PCB) implementations, on-chip solutions cannot possibly hope to attenuate noise by sprinkling several nanofarad and even microfarad capacitors across sensitive supplies, input pins, and output pins without increasing the silicon real estate to cost-prohibitive levels. Just to cite an example, consider that capacitance-per-unit areas of 15–20 fF/µm^2 are typical in state-of-the-art technologies and one 1 nF would occupy roughly 225×225 to 260×260 µm^2 of the die, which today probably represents one-quarter to one-eighth of the total area of a relatively complex system. Regrettably, increasing areas beyond, say, 2×2 mm^2 decreases the commercial appeal of the chip because the silicon wafer's die density drops, which is indicative of how many chips and how much profits each 8- or 12-in. wafer produces. Inductors, incidentally, are even more costly because air-gap inductors not only occupy considerable space, which limits inductances to maybe 40–100 nH, but also exhibit relatively poor quality factors, which is another way of saying they include considerable parasitic series resistances.

The lower filter densities and breakdown voltages that SoC, SiP, and SoP strategies impose on the design shift the burden of attenuating noise to the power-supply and signal-processing circuits of the system. Functional blocks, like phase-locked loops (PLLs), ADCs, VCOs, and DSPs, must survive more noise in the substrate and supplies and the linear and switched regulators that supply them must also generate less and more effectively suppressed switching noise. In other words, higher bandwidth and higher power-supply rejection (PSR) in supply and interface circuits are necessary to counter the effects of noise in time, before they propagate through the system. To add insult to injury, supply voltage variations must also be lower, on the range of 10–50 mV, to extend dynamic range maximally under low breakdown voltages; this is especially difficult without a large on-board capacitor in the presence of fast rising and falling load dumps—and all this with only cost-effective CMOS and maybe bipolar–CMOS (BiCMOS) technologies.

1.2 MINIATURE SOURCES

The fundamental drawback of microscale sources is that limited space constrains energy and power to miniscule levels. What is worse, technologies that store more energy unfortunately suffer from lower power densities and vice versa. To illustrate this latter point, consider that a capacitor, which responds quickly to changing loads, supplies high power-per-unit volume, but only for a short while because energy density is low. A fuel cell of equivalent dimensions, on the other hand, which incidentally requires additional time to respond, stores more energy but sources less power, as the Ragone plot of Figure 1.3 corroborates graphically [3]. For this reason, Li ions are popular in cellular phones, tablets, laptop computers, digital cameras, and other mobile products because they represent what amounts to a balanced alternative, with not only moderate energy and power densities, but also intermediate speed. Super or ultra capacitors feature comparable trade-offs, plus additional cycle life with higher leakage power. Additionally, the voltage range of super capacitors extends to zero, below the headroom limit of a circuit, under which drawing energy is less probable; this means the circuit may not leverage some of the energy stored in the capacitor.

Compromising energy, power, and speed for cost is less appealing in small applications like wireless microsensors, where operational life (i.e., energy) is as important, for example, as wireless telecommunication, which demands considerable power. As a result, complementing the relatively higher power levels that inductors,

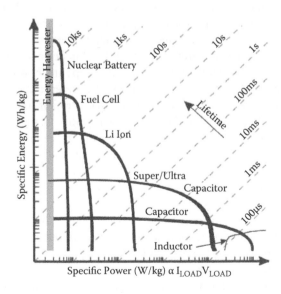

FIGURE 1.3 Ragone plot: Relative energy–power performance of various energy-storage devices.

capacitors, ultracapacitors, and Li ions supply with the energy that fuel cells can store (up to 10 times as much as Li ions can) is gathering momentum in industry and research circles alike. Miniature nuclear sources are also garnering attention, even if nuclear energy produces less power, costs more, and poses safety concerns. Energy transducers enjoy more popularity, though, because they are safe (unlike radioactive isotopes), produce few to no by-products (unlike fuel cells), and convert energy from a virtually boundless source: the surrounding environment. Ambient energy, unfortunately, is not reliable or consistent, and small transducers produce substantially low power levels. Nevertheless, the promise of extended and perhaps perpetual life is sufficient motivation to fuel research efforts in all directions.

Harnessing ambient energy from light, motion, heat, and electromagnetic radiation can certainly extend operational life, but only if power losses do not overwhelm gains. Miniature transducers and accompanying microelectronic conditioners are at the forefront of research and, at moderate scales, also at the edge of commercialization. The fact that ambient energy represents a virtually boundless source for a tiny harvester is almost sufficient motivation to drive research forward and entice industry to invest in developing relevant technologies. Lower cost and the absence of by-products further tilt the pendulum of public opinion on the side of harvesters, relegating fuel cells and atomic batteries to special applications that the military and the automobile industry, for example, might demand.

1.2.1 LIGHT ENERGY

Solar light is perhaps the most appealing source because, when exposed to the sun, photovoltaic (PV) cells can generate 10–15 mW/cm² [4], which is well above those of its counterparts. In effect, incoming photons break loosely tied electrons, as Figure 1.4 shows, and the built-in potential across the p–n junction shown in Figure 1.4(a) pulls these free electrons to the n-type region to establish current flow. The voltage that results across the cell, however, forward biases the parasitic diode that Figure 1.4(b) models to sink and lose some of the generated current.

The problem with solar energy is that output power falls drastically when direct sunlight is unavailable, down to 10–20 μW/cm². These power levels are so low that the act of harnessing, which is the work performed by circuits to transfer energy into a battery, may dissipate most, if not all of the power available, negating the harvesting objective of the system. Notwithstanding, a few researchers find solace in

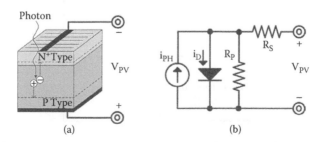

FIGURE 1.4 (a) Photovoltaic cell and (b) its corresponding electrical model.

how often microsystems idle because producing any power whatsoever for extended periods still translates to appreciable energy in the long run.

Industry and research today concentrate most of their efforts on either larger scale systems, where larger surface areas compensate for the lower power densities that artificial lighting generates, or smaller photovoltaic cells that receive direct sunlight, where even 2×2 mm^2 can generate substantial power for a microsystem. Relying on solar energy, however, limits wide-scale adoption because applications may not always place harvesting nodes in places that receive direct sunlight, so overcast days and evenings interrupt the harvesting process. Many applications, in fact, are not only indoors, like wireless microsensors in factories, hospitals, and homes, but also mobile, where the host may or may not receive sunlight, as with automobiles, bicycles, airplanes, and people. All of this is to say that harnessing artificial lighting is important, but also challenging under microscale constraints.

1.2.2   KINETIC ENERGY

By definition, mobile products, which represent an increasingly expansive market space, move. And from a commercial standpoint, harnessing kinetic energy from motion is attractive because vibrations are consistent and abundant—be it an engine at 1,000 to 10,000 revolutions per minute or a person at one to two strides per second. Kinetic transducers can also generate up to 200 µW/cm^2, providing the onboard harvesting microelectronics more power to operate than artificial lighting can. Although power is nowhere near what photovoltaic cells produce when exposed to direct sunlight, harnessing power from dependable 1–300 Hz vibrations over extended periods can accumulate appreciable energy.

Electromagnetic transducer. One way of drawing power from motion is electromagnetically [5], by allowing vibrations to move a mass attached to a conducting coil about a stationary magnet, as Figure 1.5 illustrates. The electromagnetic field that the magnet generates works against the kinetic force in the moving mass to dampen and decelerate it, thereby transferring the energy in the mass into the coil in the form of a voltage. The role of the spring is to avoid losing remnant energy by temporarily storing and releasing whatever kinetic energy remains so that the mass may once again move back toward the magnet. Unfortunately, harnessing energy this way is

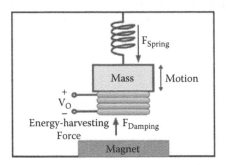

FIGURE 1.5 Transducing kinetic energy in vibrations into the electrical domain electromagnetically.

difficult in small spaces because power and voltages are substantially low at approximately 1 μW/cm^3 and tens of millivolts. Additionally, conforming and integrating a magnet and a coil into a microsystem also present considerable challenges.

Electrostatic transducer. Perhaps a more effective means of harnessing ambient kinetic energy is by pumping charge electrostatically into a storage device because the power generated is on the order of 50–100 μW/cm^2 and variable capacitors are more easily integrated with MEMS technologies. The idea is for vibrations to vary the vertical or lateral separation distance between two parallel plates of a variable capacitor C_{VAR}, like the ones that Figure 1.6(a) and 1.6(b) illustrate, so that, when a battery or another capacitor fixes the voltage across C_{VAR} to precharge voltage V_{PC}, a reduction in capacitance pumps charges q_C out of C_{VAR}:

$$C_{VAR} \equiv q_C v_C = q_C V_{PC}, \tag{1.1}$$

and generates output dq_C/dt current [6]. Alternatively, a reduction in capacitance raises C_{VAR}'s v_C when C_{VAR} is disconnected (i.e., with a fixed q_C) to augment C_{VAR}'s energy E_C:

$$E_C = 0.5 C_{VAR} v_C^2. \tag{1.2}$$

This latter scenario produces a net energy gain because, while C_{VAR} decreases linearly in E_C, v_C^2 rises quadratically. Allowing v_C to increase to, say, 100–300 V, however, which is not atypical, exceeds the breakdown limits of most standard CMOS and BiCMOS process technologies. As a result, at least for the time being, constraining v_C to voltages that near the breakdown limits of the process seems more practical than fixing q_C.

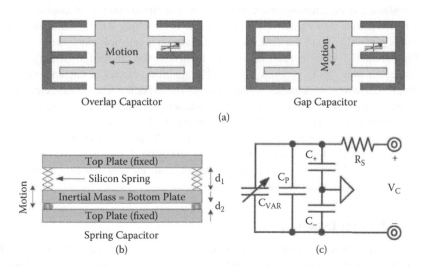

FIGURE 1.6 (a) Top and (b) vertical views of variable MEMS parallel-plate capacitors or varactors and (c) their corresponding electrical model.

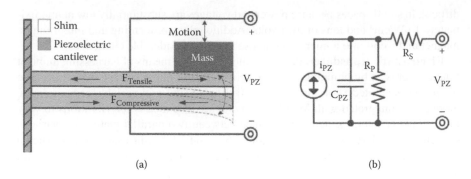

(a)   (b)

FIGURE 1.7   (a) Shimmed piezoelectric cantilever and (b) its corresponding electrical model.

The electrical equivalent of these moving parallel plates is the variable-capacitor model that Figure 1.6(c) shows. Unfortunately, the physical device also includes parasitic static capacitors C_P, C_+, and C_-. Contact material also introduces a series of parasitic resistance R_S that dissipates ohmic power. Of these, C_P, C_+, and C_- are often more problematic because they drain some of the energy that C_{VAR} harnesses from motion.

Piezoelectric transducer. Although piezoelectric transducers are perhaps more difficult to integrate, they can generate higher power, up to maybe 200 $\mu W/cm^2$. Here, when fastened to a stationary base, as Figure 1.7(a) illustrates, a shimmed piezoelectric cantilever generates electrical energy in the form of alternating charge when mechanical energy in motion induces the material to bend and oscillate about its fixed point [7]. The advantages of this approach are that piezoelectric research is relatively mature in its evolution and transducers generate higher power than their electrostatic and electromagnetic counterparts. The disadvantage, of course, is integrating the device into a small space, which is why attaching it to the outer surface of a chip is probably one of the best ways of realizing a small SoP platform.

Operationally, motion rearranges and polarizes the molecular structure of a piezoelectric material to generate charge. The resulting electrical current, which i_{PZ} in Figure 1.7(b) models, charges and discharges the capacitor C_{PZ} that appears across the terminals of the transducer. The magnitude and direction of i_{PZ} change with the acceleration and orientation of the moving mass. Unfortunately, contacts and leakage paths introduce series and parallel resistances R_S and R_P, but not to the extent that they negate the harvesting benefits of the structure. Ultimately, the energy that C_{PZ} receives is the power the transducer draws from motion, which returns to the mechanical domain when uncollected.

1.2.3   THERMAL ENERGY

Temperature represents another source of energy. Thermocouples that rely on the Seebeck effect, for example, derive thermal energy from a temperature difference.

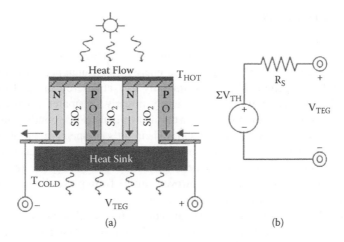

FIGURE 1.8 (a) Thermocouple transducer built out of thermal piles and (b) the corresponding electrical model.

Here, as in Figure 1.8(a), heat flux carries dominant charge carriers: electrons in n-type and holes in p-type materials, from high- to low-temperature regions, much like diffusion carries particles from high- to low-concentration regions. Here, however, departing electrons in the n-material leave behind ionized molecules in the hot side that attract the electrons that the p-region generates when holes drift toward the cold end. As these electrons flow from p- to n-type regions to deionize molecules, they harness thermal energy and jump to higher energy states. This way, electrons accumulating at the cold end of the n-type rod establish a negative potential with respect to the holes that accrue at the cold side of the p-type counterpart, generating thermally induced voltages across each pile. Since each n–p pile combination generates a small voltage v_{TH}, engineers often cascade several of these in series to produce a combined voltage v_{TEG}, which is higher. The silicon rods and interconnecting metals, of course, introduce parasitic series resistances that R_S in Figure 1.8(b) models.

Unfortunately, high temperature alone is not sufficient to generate power. Said differently, a thermocouple generates a voltage only when a temperature differential exists. Designers, as a result, usually attach one of the terminals to a heat-absorbing material. This requirement, however, impedes integration in several ways. In the first place, a heat sink occupies considerable space. In the second, finding a cool source against which to attach the transducer is not always possible. And, thirdly, output power is proportional to the temperature difference across the device and, because microsystems are tiny, temperature differentials are correspondingly low, below maybe 10°C. The voltage and power levels they produce are consequently low at about 15 μW/cm^3 [5], which nears the levels produced by photovoltaic cells when illuminated with artificial lighting and electromagnetic transducers in response to vibrations. Note that power is also proportional to the Seebeck coefficient of the thermoelectric materials used.

1.2.4 MAGNETIC ENERGY

There is also energy in a magnetic field, such as around an AC power line. Drawing power from such a source inductively, however, is challenging in three respects. To start, as in the case of harnessing kinetic energy electromagnetically, the power levels and voltages generated are substantially low at microwatts and millivolts. Unfortunately, the act of transferring energy already dissipates microwatts and, therefore, drains most of the little power that is available. Plus, conditioning power in the form of millivolts is, from an integrated circuit's perspective, problematic. As a result, some researchers opt to accumulate and convert magnetic energy into the mechanical domain first, so that a kinetic transducer can then generate power in a more benign form. Each conversion, however, suffers from power losses that further reduce what little energy was available in the first place.

The second issue with an electromagnetic source is that power falls drastically with distance: quadratically in the far field and even faster in the near field. Therefore, to produce practical power levels, the transducer should remain close to a rich source. Inductively coupling energy across a near-field separation of, say, 8 mm to charge the battery of a portable device like a cellular phone wirelessly is possible this way. However, increasing the separation beyond this point, which many remote wireless-sensor applications demand, lowers system efficiency to such an extent that extracting power is no longer practicable. The third challenge with harvesting magnetic energy is, of course, integration, because tiny inductors harness a small fraction of the magnetic energy available.

1.2.5 CONCLUSIONS

As mentioned earlier, there is no ideal source because, while harvesters, atomic batteries, and fuel cells store appreciable energy, they supply little power. Conversely, inductors and capacitors source high power, but stow little energy. And Li ions and super capacitors are moderate in both respects, sacrificing energy for power and power for energy. Additionally, while filter passives supply power almost instantaneously, Li ions and supercapacitors require time to respond—and fuel cells even more time—whereas atomic batteries and harvesters remain virtually unresponsive to load changes. As a result, systems today normally supply what is close to instantaneous power, which refers to the rising and falling edges of the load shown in Figure 1.1, with capacitors and semisustained bursts with a Li ion.

Relegating the tasks of storing energy and supplying power bursts to a Li ion or supercapacitor, however, represents a compromise in energy that is difficult to accept under microscale constraints where both energy- and power-density requirements are severe. Replacing the Li ion with a fuel cell similarly trades power for energy, loading capacitors and inductors with more power and, as a result, increasing output voltage variations in response either to load dumps or space requirements. Further decoupling energy from power is therefore important in microsystems, which is the motivation for classifying technologies into energy sources, energy/power caches, and power supplies.

Harvesters, atomic batteries, and fuel cells fall under the category of energy sources. Of these, atomic batteries are less popular because they are unsafe, and costly and tiny fuel cells because they produce chemical by-products. Li ions and supercapacitors basically cache energy and power for moderate loads, but while Li ions are perhaps more mature and popular, ultracapacitors survive more charge–discharge cycles, which is why the latter are garnering interest, albeit with higher leakage. Lastly, capacitors and inductors both supply power, but only capacitors supply instantaneous current, so systems use capacitors to store charge and supply quick load dumps and inductors to supply the collective steady-state needs of the load and momentary demands of capacitors.

Of the ambient sources discussed here, solar energy generates the most power, followed by kinetic energy when harnessed by piezoelectric or electrostatic transducers. Thermal and magnetic energy are less appealing under microscale integration because their power and voltage levels are low. Similarly, artificial lighting also generates low power levels, as does drawing kinetic energy with an electromagnetic device. As a result, some engineers favor solar photovoltaic and vibration-sensitive piezoelectric and electrostatic transducers over their competing counterparts.

1.3 CONDITIONING MICROELECTRONICS

The role of the power-supply circuit in a system is to transfer energy from a source and condition power to supply the needs of its load. Because energy is finite and especially scarce in a microsystem, the overriding measure of success, outside conditioning, is how much of the available power P_{IN} reaches the output as P_O. In other words, power lost P_{LOSS} across the system determines one of the most important parameters of a power supply: its efficiency η or

$$\eta \equiv \frac{P_O}{P_{IN}} = \frac{P_{IN} - P_O}{P_{IN}} = \frac{P_O}{P_O + P_{LOSS}}. \tag{1.3}$$

For example, if the load demands 100 μW and the power supply dissipates 50 μW to generate those 100 μW, the source supplies 150 μW and the system is 66.7% efficient. Similarly, in the case of harvesters, if a source supplies 100 μW, but the system dissipates 50 μW, the output only receives 50 μW at 50%.

Efficiency typically changes with output power P_O because losses do not vary at the same rate. At moderate to high power levels, for example, quiescent losses usually become a smaller fraction of a rising load, so efficiency is relatively high [8]—for example, when losing 75 μW to drive a 1 mW load yields 93%. Microsystems, unfortunately, normally demand lower power levels; this means that power losses represent a larger percentage of the load—for example, when losing 50 μW to drive 50 μW is the result of 50% efficiency. What is worse, microsystems normally idle and dissipate even less power—maybe 1–10 μW. Under these idling conditions, operational life is especially sensitive to power losses because the lifetime difference of a 1 μW and a 9 μW loss under a 10 μW load is close to 70%, or 1.7 times, and under a

100 µW load is 10%, or 1.1 times; dissipating less than 1 µW is extremely difficult, if not impossible, when powering alert functions and other vital blocks in the system. All of this is to say that light-load efficiency is probably one of the most difficult challenges to tackle in a microsystem, especially when the act of conditioning, in and of itself, requires work.

1.3.1 LINEAR SWITCH

Perhaps the most accurate means of conditioning power is linearly because the circuit is always alert, ready to respond to any and all variations in load and source. In other words, in addition to not generating switching noise, a linear conditioner reacts to oppose the effects of disturbances injected and coupled into its output. Admittedly, the circuit suppresses noise only up to its bandwidth [9], beyond which the system is unable to respond. Nevertheless, a linear circuit is relatively simple, so it reaps the high-bandwidth benefits that smaller devices and fewer transistors offer.

Architecturally, as Figure 1.9 shows, a controller modulates the conductance of a series switch linearly to conduct whatever current the system demands. In the case shown, p-channel metal–oxide semiconductor (MOS) field-effect transistor (FET) M_{SW} is only a sample, though a popular embodiment of the switch. Low-dropout (LDO) linear regulators, for example, use this topology and sample output voltage v_O to ensure that v_O remains near a user-defined target. To illustrate another application, sensing input current i_{IN} and regulating M_{SW}'s conductance or equivalent series resistance R_{SW} to draw as much i_{IN} as the source allows satisfy the objectives of a harvester.

Irrespective of its conditioning aim, what is ultimately important to conclude is that the circuit does not generate switching noise; current i_{IN} only flows to the output (i.e., in one direction), so input v_{IN} must exceed output v_O, and the circuit dissipates conduction power across M_{SW} as P_{SW}:

$$P_{SW} = i_{IN}\left(v_{IN} - v_O\right) \propto i_{IN(AVG)}, \qquad (1.4)$$

and quiescent power through the controller as P_C for a total power loss P_{LOSS} of

$$P_{LOSS} = P_{SW} + P_C = i_{IN}\left(v_{IN} - v_O\right) + I_Q v_{IN}, \qquad (1.5)$$

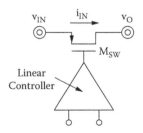

FIGURE 1.9   Basic architecture of a linear conditioner.

where $v_{IN} - v_O$ is the voltage across M_{SW} and I_Q that the current v_{IN} supplies to the controller. Although all three points are important, the last one deserves attention because microsystems cannot afford to lose power indiscriminately. Efficiency here, in fact, cannot ever reach output-to-input voltage ratio v_O/v_{IN} because the voltage across M_{SW} is $v_{IN} - v_O$ and I_Q is a finite and load-independent physical requirement for the circuit to operate properly:

$$\eta_L \equiv \frac{P_O}{P_{IN}} = \frac{P_O}{P_O + P_{LOSS}} = \frac{i_{IN}v_O}{(i_{IN} + i_Q)v_{IN}} < \frac{v_O}{v_{IN}}. \tag{1.6}$$

Consider, for example, that when a 3.6 V Li ion supplies a 1.8 V load, η_L is no better than 50%, and that is only when I_Q is negligibly smaller than i_{IN}, which is largely not the case in microsystems, particularly when they idle.

1.3.2 SWITCHED CAPACITORS

Introducing one or more intermediate steps into the energy-transfer process further decouples the needs of the input from those of the output, adding flexibility to the conditioning function. A transitional stage, therefore, receives, stores, and releases energy when prompted, as a capacitor would when configured accordingly. Charge pumps, as many circuit design engineers refer to them, switch the connectivity of "flying" capacitors in alternating phases across a switching period T_{SW} to first receive energy from a source, be it the actual input or a previous stage, and then release it to a load, which could be just another stage. Charging several flying capacitors in parallel, for instance, from input v_{IN}, as C_{F1} through C_{FN} illustrate in Figure 1.10, and discharging them in series to a load effectively "steps up" or "boosts" the input v_{IN} to a higher voltage v_O, which is an otherwise impossible feat for a linear conditioner to achieve. Similarly, just to show how flexible the architecture can be, charging capacitors from v_{IN} in series and connecting them in parallel to v_O in the alternate phase "steps down" or "bucks" v_{IN} to a lower voltage.

This flexibility, however, results at the expense of noise because decoupling v_{IN} from v_O implies that v_{IN} and v_O are, at times, disconnected, so i_{IN} momentarily raises v_{IN} and the load similarly discharges whatever output capacitance C_O is present at v_O. Allowing v_{IN} and v_O to rise and fall this way before periodically loading and recharging them with the flying capacitors creates a systematic noise ripple in the input and output. To this, the switches whose task is to reconfigure the connectivity of the network, inject additional noise because the signals that drive them at switching frequency f_{SW} rise and fall quickly across maybe 10 ns or less—coupling displacement

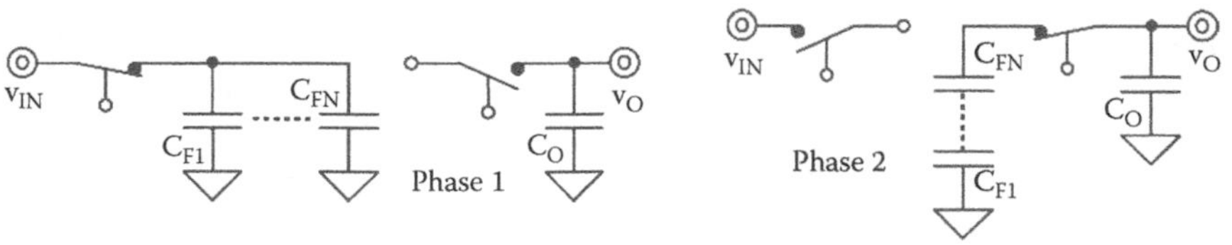

FIGURE 1.10 Switching phases of a boosting two-step switched-capacitor charge pump.

noise energy into sensitive nodes through parasitic gate–source and –drain capacitors C_{GS} and C_{DG} that complementary MOS (CMOS) switches embody.

Because the voltages across conducting switches are dynamic and decrease with time, switched-capacitor circuits do not dissipate the quasistatic and limiting $v_{IN} - v_O$ voltage-drop conduction power that linear conditioners do across M_{SW}. Still, the switches lose momentary switching losses when conducting current with decreasing, but nonetheless, finite voltages across them. The initial voltage across the switch that connects v_{IN} to the flying capacitors in phase 1 of Figure 1.10, for example, is the voltage the load drooped in phase 2, which is often not a trivial amount. Similarly, the fully charged flying capacitors and the drooped C_O also present a voltage difference to their connecting switch at the beginning of phase 2. To help quantify these losses, consider that while v_{IN} in Figure 1.11(a) loses energy E_{IN} to charge equivalent flying capacitor C_F from $v_{C(INI)}$ to v_{IN}:

$$E_{IN} = Q_C v_{IN} = C_F\left(v_{IN} - v_{C(INI)}\right)v_{IN} = C_F \Delta v_{SW} v_{IN}, \qquad (1.7)$$

C_F's energy E_C rises from $E_{C(INI)}$ to $E_{C(FIN)}$ for a positive gain of ΔE_C:

$$\Delta E_C = E_{C(FIN)} - E_{C(INI)} = 0.5 C_F v_{IN}^2 - 0.5 C_F V_{C(INI)}^2, \qquad (1.8)$$

which is nonetheless a fraction of E_{IN}, so the switch loses the difference [10] as E_{SW}:

$$E_{SW} = E_{IN} - \Delta E_C = 0.5 C_F\left(V_{IN}^2 - 2 v_{IN} v_{C(INI)} + V_{C(INI)}^2\right) = 0.5 C_F \Delta v_{SW}^2. \quad (1.9)$$

Similarly, because connecting capacitors in parallel like in Figure 1.11(b) amounts to charging them in series from a source whose voltage is the initial voltage across the switch $v_{C(INI)} - v_{O(INI)}$ or Δv_{SW}, the switch consumes

$$E_{SW} = 0.5 C_{EQ} V_{SW}^2 = 0.5\left(C_F \oplus C_O\right)\left(v_{C(INI)} - v_{O(INI)}\right)^2 \propto i_{IN(AVG)}^2 \qquad (1.10)$$

of the energy that was initially stored in C_{EQ} as $0.5\, C_{EQ} v_{C(INI)}^2$, where C_{EQ} is the series combination of C_F and C_O. In other words, charge pumps generally and necessarily lose switching energy $0.5\, C_{EQ} \Delta v_{SW}^2$ in the interconnecting switches.

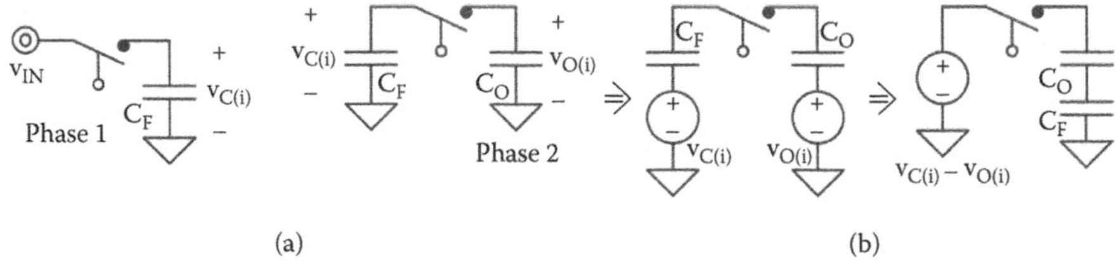

(a) (b)

FIGURE 1.11 Charging a capacitor with a switch from (a) an input source and (b) another capacitor.

Although this fundamental loss in the switch is by no means negligible, it decreases with reductions in steady-state $i_{IN(AVG)}$ because all the capacitors droop less (so Δv_{SW} is smaller) with lower currents, which is good for microsystems because they supply and demand little power. In practice, however, v_{IN} still supplies the energy switching noise and leakage currents drawn from the capacitors, the quiescent power that energizes the controller, and the power parasitic gate and base capacitors require charging and discharging every time they switch. So to summarize, a charge pump generates switching noise, can buck or boost its input, and dissipates $0.5\,C_{EQ}\Delta v_{SW}^2$ across interconnecting switches, which rises quadratically with $i_{IN(AVG)}$. Relative to its linear counterpart, the output is noisy and less accurate, but also flexible. The circuit also impresses voltages across the switches that decrease not only with load but also with time. Additionally, charge pumps dissipate power to charge and discharge parasitic capacitors that would normally not switch periodically in a linear conditioner.

1.3.3 Switched Inductor

Just as charge pumps employ capacitors to transfer energy, magnetic-based switching converters use inductors to cache and release input energy temporarily to one or several loads [11]. They achieve this by energizing and draining inductors in alternating cycles. In more specific terms, an input source v_{IN} energizes a transfer inductor L_X by inducing L_X to draw current from v_{IN}—that is, by connecting the other terminal of L_X to a lower voltage like ground in phase 1 of Figure 1.12(a). During this time, L_X's current i_L rises linearly, as the graph in Figure 1.12(b) shows, because L_X's voltage v_L or $L_X di_L/dt$ is positive and, for the most part, constant at, in this case, $v_{IN} - 0$, or more generally, at energizing voltage v_E. Reversing the polarity of v_L to, say, $-v_O$, as shown in phase 2 of Figure 1.12(a), causes L_X to release the energy it received during phase 1. Because the circuit either regulates the voltage across a load or charges a battery with a well-defined, low-ripple, slow-changing voltage, v_O is, for all practical purposes, a constant and, in this case, also L_X's de-energizing voltage v_{DE}. As a result, i_L generally rises at v_E/L_X when v_{IN} energizes L_X across phase 1 and falls at v_{DE}/L_X when L_X delivers energy to v_O in phase 2 [12].

A switched-inductor conditioner embodies a few idiosyncrasies worth mentioning. First, L_X conducts a steady-state current $i_{L(AVG)}$ about which the ripple in Figure 1.12(b) rides. Conducting current continuously like this places the circuit in what experts call continuous-conduction mode (CCM). Inducing L_X to stop conducting current

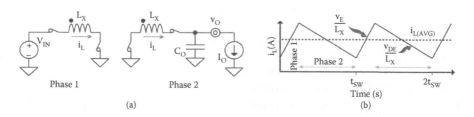

FIGURE 1.12 (a) Switching phases of a switched-inductor converter and (b) the time-domain response of its inductor current.

momentarily, which is equivalent to i_L reaching zero and remaining there for a finite fraction of the period, amounts to operating in discontinuous-conduction mode (DCM). What sets $i_{L(AVG)}$ is either the power needs of a load or the sourcing capabilities of its source.

Another trait or, rather, feature is that a switched-inductor circuit can buck or boost v_{IN} because L_X can release energy to any output voltage as long as energizing and de-energizing voltages v_E and v_{DE} remain positive and negative, respectively. In fact, permanently attaching L_X's input terminal to v_{IN} is a derivative embodiment of the general form shown in Figure 1.12(a) that, to release L_X's energy, requires v_O to exceed v_{IN}. Likewise, connecting L_X's output terminal to v_O directly demands that v_{IN} stay above v_O for L_X to energize. These two special cases implement the well-known boost [13] and buck [14] configurations reported in the literature. One last peculiarity to note is that inductors, unlike capacitors, receive or deliver energy, but do not store it statically over time; this is not a problem in conditioners because the overriding aim is to draw and deliver power—not store it, as rechargeable batteries and large capacitors would.

Within the context of operational life and efficiency, the only elements in a power stage that incur first-order conduction losses are the switches because a capacitor does not conduct steady-state current and the steady-state voltage across an inductor is zero. Interestingly, while capacitors hold the initial voltage across a switch by sourcing whatever instantaneous currents are necessary, an inductor maintains its current steadily by instantly swinging its voltage until it finds a suitable source or sink for its current. In other words, after a connecting switch opens, i_L raises or lowers L_X's open terminal voltage instantaneously until a source or sink clamps it. As a result, the small series resistance R_{SW} in the interconnecting switches and power path drop a voltage v_{SW} or $i_L R_{SW}$ that consumes power P_{SW} or $i_{L(RMS)}^2 R_{SW}$. Since R_{SW} is small, v_{SW} and P_{SW} are both low. This quasi-lossless property is the driving force behind the commercialization and general adoption of switched-inductor conditioners.

In practice, these conduction losses are small, but nevertheless finite, so they reduce efficiency, as do second-order losses similar to those found in linear and charge-pump circuits. The controller, for example, requires sufficient quiescent current to manage the system and deliver a command signal in time before the onset of the next switching cycle. The circuit also dissipates switching power to charge and discharge the parasitic capacitors that the switches present to the controller. When lightly loaded, in fact, second-order quiescent and gate-drive losses become as important as and, in some cases, more important than first-order switch losses because the voltage dropped and power lost across the switches fade with decreasing load levels [15].

Note, however, that good and reasonably sized power inductors are bulky and difficult to integrate [16]. The unfortunate truth is that on-chip inductances typically fall below 100 nH and introduce considerable parasitic equivalent-series resistances (ESRs) that further dissipate conduction power and degrade efficiency. Discrete inductors, on the other hand, are considerably better and, to a certain extent, also relatively small—for example, a $2 \times 2 \times 1$ mm^3 1 μH inductor that saturates at 1.8 A and introduces 0.2 Ω of series resistance. Accordingly, switched-inductor conditioners in microsystems should employ and co-package no more than one power inductor, which is why single-inductor multiple-output (SIMO) strategies are garnering interest in research circles [11]. Using smaller inductances is certainly possible, though at the expense of higher conduction losses because inductor ripple current Δi_L increases or gate-drive losses because switching frequency f_{SW} rises to bound Δi_L.

Because these magnetic-based circuits switch, they inject switching noise, like charge pumps do. If L_X connects directly to v_{IN} or v_O, however, as in boost or buck converters, L_X's average current $i_{L(AVG)}$ matches the source or load to keep the capacitance at v_{IN} or v_O from charging or discharging considerably. Even so, v_{IN} and v_O do not source or sink ripple currents, so Δi_L charges and discharges C_{IN} or C_O to create a corresponding ripple in v_{IN} or v_O. Unfortunately, decoupling the input or output from L_X with a switch, as in buck or boost converters, allows the source or load to charge or discharge C_{IN} or C_O when the switch opens, increasing the ripple in v_{IN} or v_O. In all, these magnetic conditioners generate switching noise, can buck or boost their inputs, dissipate $i_{L(RMS)}{}^2 R_{SW}$ across their interconnecting switches that rises quadratically with $i_{IN(AVG)}$, and, when applied to microsystems, should employ no more than one co-packaged inductor.

1.3.4 CONCLUSIONS

In comparing performance, the functionality of the conditioners precedes all other parameters, and in the case of microsystems, losses arguably follow because they limit how long a system operates. So, on the first count, switching circuits can boost their inputs while linear circuits cannot—except that a boosting function is not always necessary. Consider, for example, that while boosting a 0.4 V photovoltaic cell to charge a 2.7–4.2 V Li ion requires a switching circuit, bucking a 2.7–4.2 V Li ion to supply a 1.8 V load does not.

From the perspective of power, as summarized in Table 1.1, the switch in a linear circuit drops $v_{IN} - v_O$ continuously, whereas the terminal voltages across the switches in a charge pump and a switched inductor decrease with load and across time. In other words, switched circuits can dissipate less power than their linear counterparts. Switched capacitors, however, typically require more switches to implement than switched inductors, so capacitor networks dissipate more gate-drive losses P_{GD}.

Switch losses in all three cases, however, diminish with load, so when microsystems idle, efficiency is more sensitive to second-order losses, which is where linear

TABLE 1.1
Comparing Conditioning Circuits

	v_O	First-Order Loss P_{SW}	Second-Order Losses	Switching Noise
Linear	$< v_{IN}$	$i_{IN}(v_{IN} - v_O)$	P_Q	None
Switched C	$\leq \geq v_{IN}$	$0.5 C_{EQ}\Delta v_{SW}{}^2 f_{SW}$	$P_Q + P_{GD}$	$\propto i_{IN(AVG)}, C_{PAR}, 1/f_{SW}$
		$\Delta v_{SW} \propto i_{IN(AVG)}, t$		
Switched L	$\leq \geq v_{IN}$	$i_{L(RMS)}{}^2 R_{SW}$	$P_Q + P_{GD}$	$\propto \Delta i_L \propto 1/L_X, 1/f_{SW}$
		$i_{L(RMS)} \propto i_{IN(AVG)}$		And maybe $i_{IN(AVG)}, C_{PAR}$

circuits might gain an edge because they do not dissipate gate-drive power P_{GD}. The problem is that miniature devices demand little to no power when idling and moderate to substantial power when transmitting data wirelessly, so a linear circuit is more appealing when idling and less appealing otherwise, and vice versa for a switched inductor. In the end, functionality may become the deciding factor, which is why operating in DCM and reducing switching frequency f_{SW} in magnetic-based converters is so important when the load is light, as is lowering quiescent power P_Q in all three schemes. Notice, by the way, that parasitic series resistances in the inductor, capacitors, and board also dissipate ohmic losses.

Noise in the output may not be a problem for charging a Li ion but is certainly an issue when supplying a functional load like a data converter or sensor-interface circuit whose sensitivity to noise in the supply is often severe. Noise at the input is similarly problematic when drawing power from a PV cell because raising the voltage increases the current lost to the parasitic diode present and letting it drop lowers how much current the cell generates. Linear switches outshine their switching counterparts in this regard because they never disconnect from v_{IN} or v_O. Ripple current Δi_L in switched-inductor circuits, of course, adds noise. Charge pumps are probably the worst because flying capacitors momentarily and necessarily disconnect from v_{IN} and v_O. Switched-inductor circuits that decouple the outputs from the inductor also suffer from similar effects, in addition to the $\Delta_i L$-induced noise already mentioned. Ultimately, reducing noise amounts to raising f_{SW} (to shorten how long v_{IN} and v_O float) and circuit bandwidth (to keep up with f_{SW}), and, in the case of inductor circuits L_X (and f_{SW}), to reducing Δi_L. Increasing f_{SW}, however, by and large degrades efficiency, and raising L_X demands more space, which is the conundrum engineers eventually face when designing these types of circuits: when and how to trade accuracy, efficiency, and integration.

1.4 ENERGY-HARVESTING CHARGERS AND POWER SUPPLIES

The conditioners described in the last section form the foundation from which modern chargers and power supplies draw inspiration. Since no converter is ideal, engineers first prioritize specifications and then choose one or a combination of approaches according to the most important parameters in the system. In a cellular phone, for example, battery life is as sensitive to conversion efficiency as a large fraction of the load is to noise in the supply. In such a case, a switched-inductor converter bucks a Li ion voltage efficiently to a level that minimizes the power lost across a cascaded linear voltage regulator, whose purpose is to filter the noise that the switcher generates in the first place and supply the power that the sensitive load demands. In a harvester, saving energy is the chief objective, so an efficient switched-inductor charger might fit the bill, if power levels are sufficiently moderate for conduction losses to remain significant. Note that supplying a functional load requires a voltage source and feeding a battery a current source or charger, and while circuits may generate these without feedback, shunt- or series-sensed control loops often modulate the conductance or duty-cycle operation of the conducting switches to regulate them [17].

Space in miniature systems is so constricting that using a single sourcing technology represents a sacrifice in energy, power, or both. In these applications, complementing the energy features of ambient energy with the power-generating capabilities of Li ions and ultracapacitors offers appealing qualities that no one source can. Justifying the need for both energy and power, however, is imperative [18] because a hybrid supply is necessarily more complex than a single-source system. From this viewpoint, most microsystems, like wireless microsensors, idle and communicate information wirelessly, so their power range is vast and as critical as energy because the latter sets operational life. Hybrid supplies in tiny devices are therefore justifiable.

1.4.1 BATTERY-ASSISTED PHOTOVOLTAIC CHARGER–SUPPLY EXAMPLE

The switched inductor in Figure 1.13 draws power from both a 0.25–0.4 V PV cell v_{PV} and a 1.8 V power source v_{PS} or battery to supply a 1 V load [19]. Relying on only one small off-chip power inductor L_X is important because off-chip inductors are bulky and their on-chip counterparts are poor, but altogether excluding the inductor drastically reduces power-conversion efficiency. Input and output capacitors C_{IN} and C_O are also critical because they help keep L_X, the PV cell, and load currents i_L, i_{PV}, and i_O from slewing v_{PV} away from its maximum-power point and output v_O away from its target V_{REF}.

When lightly sourced, L_X draws and delivers energy to v_O—first from v_{PV} and then from v_{PS}. For this, switches S_{PV} and S_E energize L_X from v_{PV}, and S_{PV} and S_O subsequently drain L_X into v_O. After that, S_{PS} and S_O similarly energize L_X, and S_{DE} and S_O de-energize L_X into v_O. Otherwise, when heavily sourced, L_X supplies P_O to the load from v_{PV} and charges v_{PS} with what remains of P_{PV}. As before, S_{PV} and S_E energize L_X, and S_{PV} and S_O drain L_X into v_O, but unlike before, S_O opens after v_O receives sufficient energy to satisfy P_O, and i_L therefore charges switching node $v_{SW.O}$ until diode D_{PS} forward-biases and charges v_{PS} with i_L.

The PV cell generates the most power when v_{PV} nears its optimal value of $V_{PV(OPT)}$, which means that this system should adjust v_{PV} to $V_{PV(OPT)}$. Since the switching network induces a ripple voltage in v_{PV} that shifts the PV cell from its maximum-power point, v_{PV}'s ripple Δv_{PV} should be small. The circuit should also be able to modify and track v_{PV} to new targets to accommodate changes in light intensity and conditions.

Harvesting performance hinges on reducing power losses. Considering this, tiny PV cells generate microwatts, ensuring that L_X conducts continuously without

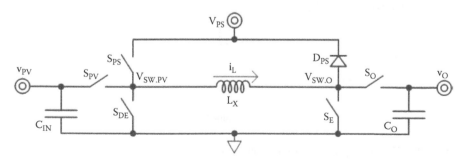

FIGURE 1.13 Battery-assisted photovoltaic buck–buck charger supply.

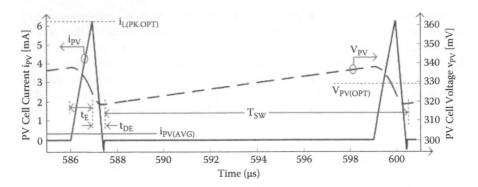

FIGURE 1.14 Photovoltaic-voltage and inductor-current time-domain waveforms.

reversing the direction of current amounts to keeping its rippling current Δi_L within a small window, which happens when L_X switches quickly. Charging and discharging the gates of power switches more often, however, requires more power, which is why L_X should not conduct continuously and should switch slowly.

To ensure that L_X derives sufficient PV power P_{PV} in one energizing (t_E) and de-energizing (t_{DE}) sequence, i_L rises and peaks to $i_{L(PK)}$ (about 6 mA in Figure 1.14). Transferring a larger energy packet E_{PV} in 1.6 µs draws sufficient P_{PV} from v_{PV} to keep v_{PV} from rising excessively over $V_{PV(OPT)}$ across the remaining 11.4 µs of the 13 µs period T_{SW}. Had E_{PV} been smaller and T_{SW} shorter, Δv_{PV} would have been less than 20 mV; however, the power lost in switching more often negates the benefits of a smaller ripple.

Selecting the channel width–length ratios of the switches that balance conduction and gate-drive losses in the system and choosing the $i_{L(PK)}$–T_{SW} combination that reduces the percentage of PV power P_{PV} lost to P_{LOSS} enable the PV cell to supply the load fully or partially and, when possible, recharge the battery. When lightly loaded, the system regulates v_O and steers excess P_{PV} to the battery to charge v_{PS} in staircase fashion, as Figure 1.15 shows. When lightly sourced, the system draws both PV and battery power to supply and regulate the load.

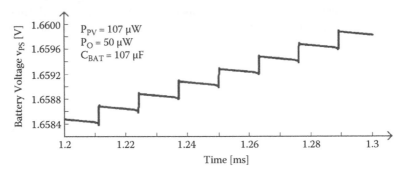

FIGURE 1.15 Battery-charge profile when heavily sourced.

1.4.2   PIEZOELECTRIC CHARGER EXAMPLE

Drawing power from an intermittent source to supply the needs of an unpredictable and uncorrelated load is challenging. As a result, harvesting and regulation are often separate and distinct functions in a microsystem. While a regulator, for example, supplies a load from a battery, a harvester can charge the battery. So, in the more specific case of tiny piezoelectric conditioners, chargers convert and condition rippling sources to replenish capacitors, rechargeable batteries, or a combination of both capacitors and batteries.

Typical battery-supplied implementations use a bridge-based rectifier to convert an alternating signal into a steady-state voltage that subsequently feeds another converter whose aim is to condition and steer power into a battery. Unfortunately, each conditioning function dissipates power, and, collectively, they diminish the gains of a microscale harvester. To minimize this loss, the piezoelectric charger in Figure 1.16 combines rectification and conditioning into one stage [20]. The underlying aim is to derive power from positive piezoelectric voltages with harvesting inductor L_H and reverse L_H's current flow to draw energy from negative voltages.

First, through v_{PZT}'s positive half cycle in Figure 1.17, switches S_I and S_N remain open to decouple the power stage from the transducer until v_{PZT} reaches its positive peak. S_I and S_N then engage (at 15.73 ms) to discharge C_{PZT} into L_H, until i_L peaks. Since LC resonance drives this energy transfer, the system estimates L_H's energizing time by waiting for one-quarter of L_H–C_{PZT}'s resonance period at $0.5\ \pi\sqrt{(L_H C_{PZT})}$. Note that sensing i_L's peak directly is more accurate, but also considerably more lossy. After this, S_N opens and i_L charges the parasitic capacitance at switching node v_{SW}^+ quickly until noninverting diode-switch D_N forward biases and depletes L_H into V_{BAT}. During the negative half cycle, S_I and S_N again open to disconnect the circuit from the transducer until v_{PZT} reaches its negative peak. Afterward, S_I and S_N discharge C_{PZT} into L_H (at 20.69 ms) for one-quarter of L_H–C_{PZT}'s resonance period. S_I then opens and D_I conducts i_L into V_{BAT}. This way, the system deposits energy into V_{BAT} every half cycle in staircase fashion, as Figure 1.18 illustrates under vibrations of various strengths.

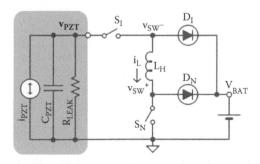

FIGURE 1.16   Single-inductor energy-harvesting piezoelectric charger.

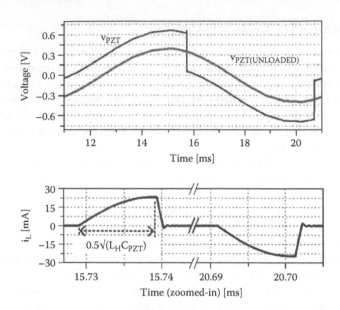

FIGURE 1.17   Time-domain waveforms.

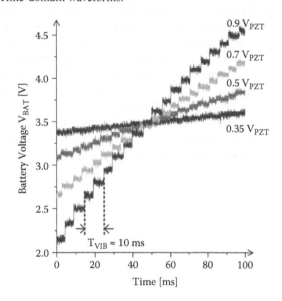

FIGURE 1.18   Battery-charge profile.

1.4.3 ELECTROSTATIC CHARGER EXAMPLE

Harnessing energy from a variable capacitor C_{VAR} is possible because the work that vibrations exert to separate the conducting plates decreases the device's capacitance. As a result, because charge q_C is the product of C_{VAR} and its voltage v_C or $C_{VAR}v_C$, reducing C_{VAR} raises v_C, and the quadratic gain that v_C^2 represents in energy $0.5\ C_{VAR}v_C^2$

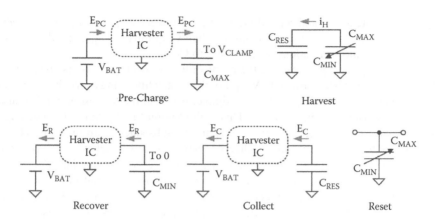

FIGURE 1.19   Electrostatic harvesting sequence.

overwhelms C_{VAR}'s linear drop. Unfortunately, v_C often rises to 100–300 V, well beyond the breakdown limits of low-cost CMOS process technologies.

Alternatively, clamping C_{VAR} to a large reservoir capacitor C_{RES} precharged to V_{CLAMP} constrains v_C and causes q_C to fall in response to reductions in C_{VAR}, which means that C_{VAR} drives charge into C_{RES}. The cycle-by-cycle progression shown in Figure 1.19 does just this: first, precharges C_{VAR} to V_{CLAMP} and then connects C_{VAR} to C_{RES} so that reductions in C_{VAR} pump charge into C_{RES}. Afterward, the system disconnects C_{VAR}, recovers remnant energy in C_{VAR}, and collects harvested charge into C_{RES}. After vibrations finish resetting C_{VAR} to C_{MAX}, the system starts another harvesting cycle by again precharging C_{VAR} to V_{CLAMP} [21].

Unfortunately, breakdown voltages in modern CMOS process technologies are low at 1.8–5 V. This means that the energy C_{VAR} holds at C_{MIN} with V_{CLAMP} can be so low that conduction and gate-drive losses in the recovery process can dissipate most, if not all of C_{VAR}'s energy. In these cases, skipping the recovery phase and clamping C_{VAR} directly to V_{BAT} (instead of C_{RES}) during the harvesting phase are prudent design choices because they reduce the sequence to precharge, harvest (to V_{BAT}), and reset and eliminate lossy energy-transfer transactions between C_{RES} and V_{BAT}. The switching network of Figure 1.20 does just this: *Precharge* C_{VAR} at C_{MAX} from

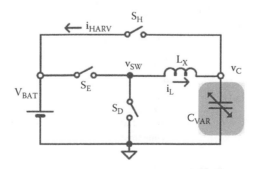

FIGURE 1.20   Battery-clamped electrostatic harvesting charger.

V_{BAT} with switches S_E and S_D and inductor L_X, connect C_{VAR} to V_{BAT} and *harvest* C_{VAR}'s charge directly to V_{BAT} through S_H when C_{VAR} falls to C_{MIN}, and disconnect all switches to allow vibrations to *reset* C_{VAR} to C_{MAX}.

In this simplified progression, there are only two energy transfers: when precharging C_{VAR} and harvesting into V_{BAT}. For the first, while using a switch to connect and precharge C_{VAR} at C_{MAX} to V_{BAT} is simple, it is also the lossiest option because the switch consumes $0.5C_{MAX}V_{BAT}^2$. This is the reason that S_E starts the precharge phase by energizing inductor L_X from V_{BAT} and S_D subsequently drains L_X into C_{VAR} [22]. Notice that C_{VAR} also charges when L_X energizes, so L_X caches only part of the energy that C_{VAR} ultimately receives when L_X finishes de-energizing into C_{VAR}.

Since C_{VAR} precharges to V_{BAT} (at 0 and every 33 ms after that in Figure 1.21), connecting C_{VAR} to V_{BAT} after precharge with S_H draws no current and, therefore, no power from either C_{VAR} or V_{BAT}. Afterward, through the first 16.5 ms of every cycle in Figure 1.21, C_{VAR} falls to C_{MIN} to drive harvesting current i_{HARV} into V_{BAT} via S_H. Because i_{HARV} is low, S_H dissipates little power across this phase. When C_{VAR} reaches C_{MIN} (at 16.5 and every 33 ms after that in Figure 1.21), all switches disconnect and, through the ensuing 16.5 ms, vibrations separate C_{VAR}'s plates until C_{VAR} rises to C_{MAX}, at which point another cycle begins.

As with every harvesting system, the ultimate goal is to generate sufficient power to output a net energy gain. The challenge is that miniature transducers generate little power and conditioning circuits can easily dissipate much of that power. In the electrostatic case presented, V_{BAT} invests $0.5\,C_{MAX}V_{BAT}^2$ as E_{INV} to charge C_{VAR} to V_{BAT} at C_{MAX} and harvests energy from C_{VAR} as C_{VAR} falls to C_{MIN}. As a result, V_{BAT} loses energy when the cycle begins (at 0 and every 33 ms after that in Figure 1.21) and harvests energy every half cycle after that to produce the staircase response

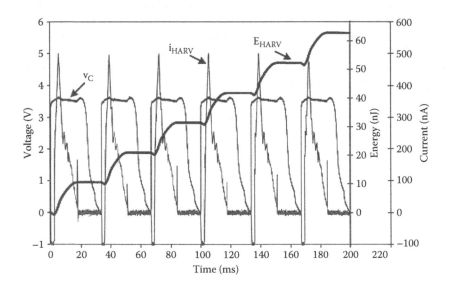

FIGURE 1.21   Battery-constrained electrostatic harvesting charger.

in Figure 1.21. In other words, C_{VAR} generates more energy E_{OUT} than the system invests in E_{INV} and dissipates in E_{LOSS}:

$$E_{HARV} = E_{OUT} - E_{INV} - E_{LOSS}$$

$$= \Delta C_{VAR} V_{BAT}{}^2 - 0.5 C_{MAX} V_{BAT}{}^2 - E_{LOSS} \qquad (1.11)$$

$$= 0.5 C_{MAX} V_{BAT}{}^2 - C_{MIN} V_{BAT}{}^2 - E_{LOSS},$$

where E_{OUT} is the charge energy that V_{BAT} receives as $\Delta Q_C V_{BAT}$ or $(\Delta C_{VAR} V_{BAT}) V_{BAT}$ when C_{VAR} falls to C_{MIN} and ΔC_{VAR} is the difference between C_{MAX} and C_{MIN}. An important observation to note here is that output power from E_{HARV} is ultimately higher with larger C_{MAX}–C_{MIN} spreads and lower power-conditioning losses.

1.4.4 CONCLUSIONS

Hybrid supplies in microsystems generally suffer from full-scale integration and poor low-power efficiency. Tiny spaces, to start, constrain energy to such an extent that efficiency concerns overwhelm all other performance metrics, which means switched-inductor conditioners become almost indispensable. Power inductors, unfortunately, are bulky and difficult to integrate; this means that miniaturized chargers and supplies cannot afford to use more than one. And even if they are more efficient than linear and switched-capacitor implementations under moderate to heavy loads, magnetic-based supplies still dissipate conduction, gate-drive, and quiescent power. In fact, a fundamental challenge that designers face in ultra low-power chargers and supplies is keeping gate-drive and quiescent losses low—that is, managing complex systems with little to no power.

Harvesters bear the additional burden of synchronizing circuits to often unpredictable and uncorrelated ambient events and conditions. Not only that, starting and operating a system with no initial energy is especially problematic. Many implementations, therefore, rely on a battery storing sufficient initial charge to bias and start the system, much like cars depend on charged batteries to ignite their engines. In other words, ultrasmall solutions often complement a harvesting source with another technology, like a battery with some charge, to ensure that the system starts properly.

1.5 SUMMARY

The potential market space for miniature systems like wireless microsensors is vast. One contributing factor is that adding intelligence to expensive and difficult-to-replace infrastructures nonintrusively has the potential of updating and improving otherwise obsolete and archaic technologies of scale. Power grids can manage not only energy more efficiently with embedded instrumentation but also military base camps, space stations, industrial plants, hospitals, farms, and others. Biomedical implants and portable and wearable consumer electronics also reap the benefits of microscale integration. Tiny spaces, however, limit energy and power, and modern mixed-signal systems demand both: energy for operational life and power for functionality, like sensing, data processing, and wireless transmission.

Sadly, no single miniature source is ideal. While capacitors and inductors source high power, for example, they store little energy. Fuel cells and nuclear batteries may store more energy, but they supply considerably less power. Li ions and ultra-capacitors are moderate in almost every way; however, sacrificing energy for power, or vice versa, for the sake of adopting only one technology (for low cost) is quickly becoming less viable under microscale constraints. From this perspective, harvesting ambient energy is a good way of replenishing and complementing a power-dense device that is easily exhaustible. In fact, harvesters do not release the by-products that fuel cells do, nor do they impose the safety hazards or cost infrastructure that nuclear batteries do.

Harnessing solar energy generates the most power at several milliwatts per square centimeter; however, direct sunlight is not always available and artificial lighting outputs substantially lower power levels on the microwatt per square centimeter range. Unfortunately, deriving power from thermal and magnetic sources in small spaces not only produces millivolt signals that are difficult to condition but also generates only microwatts per cubic centimeter. Although kinetic energy in motion cannot generate the milliwatts that photovoltaic cells can when exposed to solar light, vibrations are abundant and consistent, and piezoelectric and electrostatic transducers are capable of producing moderate power levels from vibrations at around 100–300 μW/cm^3. All the same, research continues and the final verdict on the power capabilities of light, thermal, magnetic, and kinetic transducers is far from certain.

Irrespective of the source, conditioning microwatts is difficult because the mere act of transferring energy can dissipate much of the little power that is available. Linear converters, for example, may be noise free, simple, and relatively fast, but their conducting switches dissipate ohmic losses that only decrease linearly with output current. And while losses in switched capacitors decrease quadratically with load, switched inductors, whose losses also decrease at a similar rate, employ considerably fewer switches, so gate-drive losses are corresponding low. At microwatts, though, quiescent and switching losses become more significant, obscuring the boundary regions where one conditioner's efficiency clearly outperforms that of another.

Mixed-signal microsystems often reduce losses and increase integration densities by mixing multiple sourcing technologies with magnetic-based converters that use only one inductor. Irrespective of the combination, hybrid supply circuits must carefully manage energy and power flow for maximum life and accurately condition power to supply a load and/or replenish a battery fully. Directing how and where power flows from a tiny photovoltaic cell with only a single inductor, for example, is as important as charging a co-packaged thin-film Li ion and regulating the voltage across a load. How circuits load transducers, charge batteries, and synchronize to ambient conditions eventually sets the operational life of a system. Designers must therefore implement these and other basic functions without exhausting or stressing the tiny sources in the microsystem; in other words, engineers must design circuits that perform relatively complex tasks with little to no energy.

REFERENCES

1. G. A. Rincón-Mora, and M. Chen. 2005. Self-powered chips—The work of fiction. *Power Management Design Line (PMDL)*. April 28, 2005 (http://www.eetimes.com/design/power-management-design/4011558/Self-powered-chips—The-work-of-fiction).
2. B. Sahu, and G. A. Rincón-Mora. 2004. A high-efficiency linear RF power amplifier with a power-tracking dynamically adaptive buck-boost supply. *IEEE Transactions on Microwave Theory and Techniques* 52 (1): 112–120.
3. M. Chen, J. P. Vogt, and G. A. Rincón-Mora. 2007. Design methodology of a hybrid micro-scale fuel cell-thin-film lithium ion source. *IEEE International Midwest Symposium on Circuits and Systems (MWSCAS)*, Montreal, Canada, August 5–8, 2007.
4. E. O. Torres, and G. A. Rincón-Mora. 2008. Energy-harvesting system-in-package (SiP) microsystem. *ASCE Journal of Energy Engineering* 134 (4): 121–129.
5. E. O. Torres, and G. A. Rincón-Mora. 2005. Long lasting, self-sustaining, and energy-harvesting system-in-package (SiP) sensor solution. *International Conference on Energy, Environment, and Disasters (INCEED)*, Session A-2, ID 368, pp. 1–33, Charlotte, NC, July 2005.
6. E. O. Torres, and G. A. Rincón-Mora. 2009. Electrostatic energy-harvesting and battery-charging CMOS system prototype. *IEEE Transactions on Circuits and Systems (TCAS) I* 56 (9): 1938–1948.
7. D. Kwon, and G. A. Rincón-Mora. 2009. A rectifier-free piezoelectric energy harvester circuit. *IEEE International Symposium on Circuits and Systems (ISCAS)*, Taipei, Taiwan, May 24–27, 2009.
8. G. A. Rincón-Mora. 2009. *Power IC design—From the ground up*. Blackfoot, ID: Rocky Mountain Publishing.
9. V. Gupta, G. A. Rincón-Mora, and P. Raha. 2004. Analysis and design of monolithic, high PSR, linear regulators for SoC applications. *Proceedings of IEEE International System on Chip (SOC) Conference*, 311–315, Santa Clara, CA.
10. G. A. Rincón-Mora. 2005. *Power management ICs: A top-down design approach*. Raleigh: Lulu.com.
11. D. Kwon, and G. A. Rincón-Mora. 2009. Single-inductor multiple-output (SIMO) switching DC–DC converters. *IEEE Transactions on Circuits and Systems II (TCAS II)* 56 (8).
12. D. Kwon, and G. A. Rincón-Mora. 2009. Operation-based signal-flow AC analysis of switching DC–DC converters in CCM and DCM. *IEEE International Midwest Symposium on Circuits and Systems (MWSCAS)*, Cancún, Mexico, August 2–5, 2009.
13. N. Keskar, and G. A. Rincón-Mora. 2008. A fast, sigma-delta boost DC–DC converter tolerant to wide LC filter variations. *IEEE Transactions on Circuits and Systems (TCAS) II* 55:198–202, February 2008.
14. H. P. Forghani-zadeh, and G. A. Rincón-Mora. 2007. An accurate, continuous, and lossless self-learning CMOS current-sensing scheme for inductor-based DC–DC converters. *IEEE Journal of Solid-State Circuits* 42 (3): 665–679, March.
15. S. Kim, and G. A. Rincón-Mora. 2009. Achieving high efficiency under micro-watt loads with switching buck DC–DC converters. *Journal of Low Power Electronics (JOLPE)* 5 (2): 229–240.
16. G. A. Rincón-Mora, and L. A. Milner. 2005. How to fully integrate switching DC–DC supplies with inductor multipliers. *Planet Analog*, eetimes.com, December 18, 2005.
17. G. A. Rincón-Mora. 2009. Analog IC design: An intuitive approach. Raleigh: Lulu.com.

18. M. Chen, J. P. Vogt, and G. A. Rincón-Mora. 2007. Design methodology of a hybrid micro-scale fuel cell-thin-film lithium ion source. *IEEE International Midwest Symposium on Circuits and Systems (MWSCAS),* Montreal, Canada, August 5–8.

19. R. Damodaran, and G. A. Rincón-Mora. 2013. Battery-assisted and photovoltaic-sourced switched-inductor CMOS harvesting charger–supply. *IEEE's International Symposium on Circuits and Systems (ISCAS),* Beijing, China, May 19–23.

20. D. Kwon, and G. A. Rincón-Mora. 2010. A 2 μm BiCMOS rectifier-free AC–DC piezo-electric energy harvester-charger IC. *IEEE Transactions on Biomedical Circuits and Systems (TBioCAS)* 4 (6): 400–409.

21. E. Torres, and G. A. Rincón-Mora. 2009. Energy budget and high-gain strategies for voltage-constrained electrostatic harvesters. *IEEE International Symposium on Circuits and Systems (ISCAS),* Taipei, Taiwan, May 24–27, 2009.

22. E. O. Torres, and G. A. Rincón-Mora. 2009. Electrostatic energy-harvesting and battery-charging CMOS system prototype. *IEEE Transactions on Circuits and Systems (TCAS) I* 56 (9): 1938–1948.

2 Low-Power Energy Harvesting Solutions for Biomedical Devices

Jordi Colomer-Farrarons, Pere Ll. Miribel-Català, E. Juanola-Feliu, and J. Samitier

CONTENTS

2.1 INTRODUCTION

Interest in the recovery of available energy from the environment is constantly increasing. The greenhouse effect in relation to climate change is a present and future concern. The possibility of converting energy available in the human environment and the capacity to transform it into useful energy has resulted in the creation of infrastructures capable of recovering tens of megawatts in the form of electrical energy—thanks, for example, to wind turbines or wave energy systems. In this way we can speak of energy harvesting on the macroscale, as is perhaps best known to the general public, but the importance of energy recovery on the microscale is even greater. What does one understand by recovery on the microscale? We should think of energy recovery that can vary in the range of nanowatts to milliwatts [1]. Evidently, approximation to these levels of recovery and interest in it depend largely on the field of application. The clearest example is in the field of portable consumer products. The possibility of keeping a mobile telephone charged without the need for a battery, with all the attendant environmental advantages, gives us a clear perspective. If electronic equipment generally could avoid the need for a battery, being charged instead by the user's own movements or by the difference in temperature between it and the environment, the existing ambient light, or the electromagnetic waves it produces in the environment, and if we extend this to millions of mobile telephones, we can extrapolate the enormous advantages of being able to develop efficient electronic systems without batteries.

In existing applications, we can speak of small-scale recovery when we refer to intelligent sensors capable of self-charging and able to transmit from the field to a remote center of communication. The concept and development of these smart wireless sensors is referred to in the present chapter. The fields of application are broad, as we will see, from the monitoring of the structural condition of bridges and highways [2] to the level of so-called intelligent buildings [3]. In general, the conception and development of intelligent and autonomous sensors that do not require the use of batteries bring many advantages from the point of view of cost, maintenance [4], etc. But the investigation and conception of these self-powered units also brings us to the field of medical healthcare in various aspects, such as the remote monitoring of chronic illnesses; avoiding continuous hospital admissions is a notable aspect [5], known as pervasive healthcare [6]. This chapter will review the principal sources of energy present in the environment that are capable of recovery and conversion to electrical energy, with particular attention to those orientated to the microscale or the nanoscale.

2.2 ENERGY HARVESTING

2.2.1 TYPES OF ENERGY HARVESTING SOURCES AND POWER RANGES

There exist various sources for the recovery of energy present in the environment, and those sources as well as the type of conducting element determine the admissible levels of recoverable energy, as do the fields of application. Among the typical sources and energy transducers available in the environment there are, for example, vibrations [7], heat [8], light [9], and radio waves [10]. The available energy per unit area, or volume, for each one of these sources depends heavily on the size, operating conditions, and technologies available [1]. Table 2.1 introduces some values for vibrational sources, ambient light, radio frequency (RF) and thermoelectrical sources. Vibrational energy is a typical approach to harvesting energy. Three main solutions are developed based on the type of transducer or microgenerator: piezoelectric ($\sim$200 μW/cm^3), electrostatic ($\sim$50–100 μW/cm^3), and electromagnetic ($<$1 μW/cm^3). There are wide ranging scenarios where this type of harvesting is or could be applied basically to monitor structural integrities. We can think of different fields of application in terms of the following mapping:

a. Stiff structures that themselves produce movement (ships, containers, mobile devices, housings of fans, escalators and elevators in public places, appliances, refrigerators, bridges, automobiles, building structures, trains)
b. Elastic structures that show an elastic deformation of their walls (rotor blades, wind mill blades, aircraft wings, pumps, motors, HVAC ducts, rotorcraft)
c. Soft structures with very low elastic modulus and high deformation ratios (different textiles, leather, rubber membranes, piping with internal fluid flow)

TABLE 2.1
Mapping of Available Energy Sources

Energy Source	Performance	
Ambient light	Indoor: 10 µW/cm^2 (low illumination)	Solar cells (6830 lx 10 W/m^2)
	Typical office: 100 µW/cm^2	Indoor solar cells (10 to 1400 lx)
	Outdoor: 10 mW/cm^2	
	Full, bright sun: 10 mW/cm^2	
Vibrational	4 µW/cm^3 (human motion—hertz range)	Microgenerators [11,12]
	800 µW/cm^3 (machines—kilohertz ranges)	350 µW [13]
	These numbers depend heavily on size,	22 µW [14]
	excitations, technologies, etc.	400 µW [15]
	Typically:	
	Piezoelectric ~ 200 µW/cm^3	
	Electrostatic ~ 50–100 µW/cm^3	
	Electromagnetic < 1 µW/cm^3	
RF	GSM: 4 µW/cm^3	<1 µW/cm^2 transmitter [12]
	WiFi: 1 µW/cm^3	~1 mW for proximate stations
	These numbers depend heavily on	(inductive coils)
	frequency of operation and distance	At 900 MHz, 1.1 m at 24.98 dBm
	between base station and receiver.	(0.315 W), ~20 µW [17]
		At 4 MHz, 25 mm (subcutaneous
		powering) > 5 mW [16]
Temperature	Human: 25–60 µW/cm^2	Thermoelectric generators
	Industry: 10 mW/cm^2	60 µW/cm^2 Thermolife ΔT = 5°C [18]

Source: Derived from Colomer-Farrarons, J. et al. *IEEE Transactions on Industrial Electronics* 58 (9): 4250–4263.

For industrial applications, it is possible to take advantage of "fixed frequency" vibrations because AC-driven motors and pumps produce vibration harmonics from their drive frequency (for example, 60 Hz in the United States and 50 Hz in Europe). For vehicles, the result is more *random* vibrations than in the other categories. Although nearly all vehicular applications provide significant vibration amplitudes and sufficient levels for energy harvesting, this energy is available more through random occurrences such as bumps, rough surfaces, and dynamic frequencies than through one particular frequency of interest.

Different industrial solutions exist in the market and some examples are introduced. Advanced Linear Devices® has developed different energy harvesting modules. EnOcean GmbH defines self-powered sensor networks and monitors energy harvesting from natural motion; MicroStrain®, the Volture system from Mide Technology®, and the Perpetuum PMG® series harvest the vibrations in their installed environment to generate electricity for wireless sensor and other applications.

Thermal generators are based on the Seebeck effect and convert the temperature difference or heat into an electrical energy. An interesting example is the thermoelectric generator (LPTG) by Thermo Life (Thermo Life Energy Corp, California) [18], measuring 0.5 cm^2 and 1.6 mm thick, that can supply energy to a biosensor when in contact with the skin.

Interesting research has been presented by IMEC [19], which has demonstrated the integration of a wireless autonomous sensor system in clothes. The system is fully autonomous for its entire life and requires no service—such as replacing or recharging the battery—from the user. The shirt with integrated electronics can be washed in a regular washing machine. The device is powered from a rechargeable battery. The battery is constantly recharged, mainly through thermoelectric conversion of the wearer's body heat. The thermoelectric generator is divided into 14 modules to guarantee user comfort. It occupies less than 1.5% of the shirt area and typically generates a power of 0.8–1 mW at about 1 V at regular sedentary office activity. However, if the user walks indoors, the power increases up to 2.7 mW at 22°C due to forced convection. The thermoelectric generator is neither cold nor intrusive for the user. In colder environments where other clothes need to be worn on top of the shirt, the power generation is typically not affected. In summary, this is one of the major fields with a huge rate of growth expected in the following years.

Ambient light is another typical example. In Nasiri, Zabalawi, and Mandic [20], the utilization of indoor cells for extremely low conditions of illumination and low voltages of operation is shown for a cell of 55 mm × 20 mm × 1.1 mm with a power of 5 µW at 10 lx to 200 µW at 1450 lx for a voltage drop of 2 V. RF harvesting is also applied, but depends heavily on frequency of operation and distance between the base station and the receiver, where the energy conversion takes place. Typical values for standard bands are 4 µW/cm^2 (GSM [global system for mobile communications]) and 1 µW/cm^2 (WiFi). In the case of inductive coils, an interesting case is reported in Carrara et al. [16], where a complementary metal–oxide semiconductor (CMOS) (solution is implemented in a 0.18 µm technology), operating with an input carrier of 900 MHz and an input voltage of 250 mV in the integrated antenna in the tag, will theoretically generate an average power of 25 µW, with an output voltage of 1.8 V to supply the tag electronics.

Attention is especially focused on body harvesting. The technological evolution is defining a new scenario in which it will be possible to monitor patients anywhere at any time (see Figure 2.1), which is of increasing interest. The traditional approach, where patients are monitored during hospital or surgery visits, would be replaced by continuous and remote monitoring, which could have a great impact on patients' quality of life. This is the concept of pervasive monitoring [21,22]. Different scenarios can be envisaged in the continuous search to meet technological challenges through miniaturization, intelligence, and autonomy of the biomedical devices [23], looking for new implantable devices or capsules, like ultrasound pill cameras. Here resides the importance of the autonomy of the devices and the importance of body harvesting, which is addressed in more detail in Section 2.2.2, along with new approaches. Some examples are introduced.

The conversion, for instance, of the natural motion of the body based on mechanical (vibration) energy to electrical energy is one of the main research topics. In this context, the power that can be harvested when a human being walks or runs has been studied at different locations, with an average of 0.5 mW/cm^3 for hip, chest,

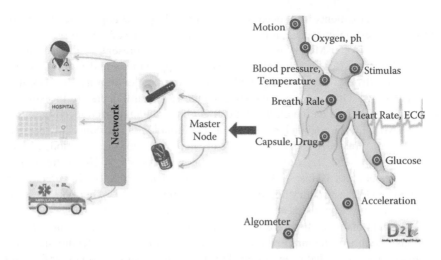

FIGURE 2.1 Typical full wireless body sensor networks (WBSNs). (From Yang, G.-Z. 2006. *Body Sensor Networks*. London: Springer. With permission.)

elbow, upper arm, and head, and a maximum of 10 mW/cm^3 for the ankle and knee. The mechanical energy can be converted to electrical energy based, for instance, on electromagnetic, electrostatic, or piezoelectric principles. But following these approaches just a few microwatts are available [23]. The other classical source of harvesting is body heat. Taking into account that the total amount of power that is waste in the form of heat is around 100 W, the energy that can be recovered, typically using the Carnot engine, is in the range of 2.4 to 4.8 W, but just a few milliwatts are really available [22].

Furthermore, in the function of the field and the application, if the quantity of energy available for recovery is small, it is useful to be able to combine different sources at the same time. In this way they are combined and at the same time the system does not depend entirely on one source. A platform that combines different sources [24] present in the environment to charge a microsystem is presented, in a form that could, at the conceptual level, be autonomous. The combination of four sources is envisaged [1]: the vibrational, based on the use of a piezoelectric generator as the conducting element; light, based on the use of interior solar cells; RF signals, through electromagnetic coupling between inductors; and, finally, a thermoelectric element. This system is implemented and described in more detail in Section 2.3 of this chapter. The ability to recover energy on the micro- or nanoscale requires the development of efficient conduction elements and this is a topic of great interest [25] in which the great importance of energy modeling in the design of these new self-powered nanodevices and the conception of nanonetworks is envisaged [26]. Recent references exist in the field of MEMS (microelectromechanical sensors) as much on the aspect of the sensor as of the energy conductor. In Khoshnoud and de Silva [27], a full review of new approaches is revised by the authors. Special interest is focused on the field of energy harvesting for self-powered sensors and their expected impact against the greenhouse effect [28].

Recent developments have been published regarding new approaches for micro- and nanoharvesting purposes in the field of transducers based on the piezoelectric effect [29]. In Trolier-McKinstry et al. [30], the authors present a report on piezoelectric thin films, looking for ultra low-voltage applications and low frequencies of operation (from 50 to 500 Hz), with power densities as high as 10^{-4} W/cm^2 at 1 g acceleration. Other energy harvesting sources and transducers are investigated. Some solutions based on the electrostatic effect have been reported recently. In Liu, Lye, and Miao [31], a sandwich parallel-plate combination of two capacitors is implemented, with a final size of 1 cm^2, capable of operating at low accelerations and low frequencies, up to 50 Hz, with a maximum recovered power of 20 nW for a maximum acceleration of 10 ms^{-2}. This is a technological approach, but in Peterson and Rincon-Mora [32] a circuitry approach to improve the recovered energy is presented that lowers conduction losses and increases the voltage across the plates.

In the field of light, there are new technological approaches [33] and circuitry solutions to improve the maximum power point tracking (MPPT) technique, which maximizes the power that is harvested. This technique represents an excessive overhead of power when it is applied to systems recovering energy at the microscale. Then, it is necessary to develop MPPT techniques with lower power consumption, as in Lu et al. [34], where from an available power of 140 µW, it is possible to recover 46.65 µW: an efficiency of 33.15%—and with power consumption, in average lighting conditions, from 784 to 2152 lx, of 180 µW, and efficiencies up to 39.51%.

New technological advances in nanomaterials present new opportunities in the future development of the storage element, with special interest in grapheme because it presents high electrical conductivity and mechanical stability, which are ideal for new storage elements, such as supercapacitors [35]. At the nanoscale, there is interesting work regarding nano-antennas, which are dipole antennas of 60 nm of width, operating at the terahertz range, where small MOM (metal–oxide–metal) tunneling diodes of 50 nm of range are envisaged for the AC rectification to recover energy from the incoming signal [36].Special interest, at the micro- and nanoscale, is focused on thermoelectric generators (TEGs), for energy harvesting purposes. In Paul et al. [37], Si/SiGe heterostructure nanoscale materials, grown on Si substrates, are derived to improve the thermoelectric performances at room temperature, improving the Seebeck coefficient compared with a bulk p-type Ge generator, with comparable doping density.

In general, at the level of the implementation of the electronics associated with these systems, they must present a trade-off between their level of intelligence and the level of energy they can recover, and their own consumption. For instance, in the case of TEG, a power management unit (PMU), which generates a DC voltage level suitable for the rest of the electronics [38], is necessary and loads present in a self-powered sensor node. A PMU [28] capable of setting up a low-input voltage from the TEG—as low as 150 mW to an output voltage for ultra low-voltage circuitries of 0.85–0.90 V is presented. In the case of vibrations coming from piezoelectric or electrostatic transducers, the PMU unit presents a sensible circuitry, which is defined by the integrated rectifier, as in the case of RF harvesting [36]. There is an interesting possibility of combining full-wave active rectifiers [39] capable of working with low-voltage amplitudes and with high efficiencies up to 90% with a charge-pump module to work with scavenged low voltages and generate regulated DC levels [40]. The charge pump, as a DC–DC converter, is designed to track the maximum power point transfer, in the range of 10 to 200 µW, for specific load conditions, operating from 0.5 to 2.5 V with efficiency close to 50%.

A more recent work by the same authors presents a different PMU with higher efficiencies, up to 93% with an MPPT also implemented, and a greater input voltage range of operation [41], from 0.44 to 4.15 V. In this case, the active rectifier also acts as voltage doubler. In the field of RF harvesting, a key point is the design also of the ACDC stage [42]. A very interesting setup is presented in Frizzell-Makowski et al. [43], where a full battery-free dynamic on/off time (DOOT) control, with a buck converter, is implemented, with efficiencies up to 95% and a low-power static consumption, as low as 217 nW.

2.2.2 FIELDS OF APPLICATION: FROM BRIDGE MONITORING TO MEDICAL IMPLANTS

We have broadly introduced the types of sources that can recovery energy from the environment, and among them we have introduced an application of interest for monitoring the mechanical condition of infrastructures (e.g., bridges [2]) in the application of sensors in the environment of intelligent buildings [3], surveillance systems [43], and medical implants [22,23]. The solutions implemented extend:

- From the use of indoor solar cells [44], from the light present in the environment, for application in control of the environment in buildings in which the low consumption of the electronics would permit the system to operate with only 100 lx to solutions in which the possibility of using electromagnetic energy present is the environment is analyzed [45]
- From the electrical network, such as radiation from radio stations for mobile telephones, to more revolutionary solutions such as those found in Zhu et al. [46], in that of the automatic sensor based on a miniature airflow energy harvester generator that consists of a wing attached to a cantilever capable of generating up to 90 µW in conditions of airflow speed as low as 2 ms⁻¹

An interesting work on very small self-powered sensors is reported in Lee et al. [47]. This is a cubic millimeter approach rather than bulky solutions, greater than 1 cm³.

But particular interest is focused in self-powered devices destined for medical applications. In the last 5 years, interest in the development of so-called body sensor networks (BSNs) has increased [5]. The miniaturization of the main electronic systems involved in such systems (as depicted in Figure 2.2), such as the instrumentation, the signal processing module, communications devices, and sensors), has opened up the increasing evolution of this field, where not only single wearable sensors are conceived, but also implantable solutions in the conception of e-health, where new trends are introduced into the medical industry. The traditional approach, where patients

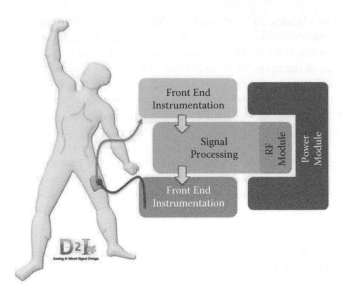

FIGURE 2.2   Miniaturization of the main involved electronics.

are monitored during hospital or surgery visits, would be replaced by continuous and remote monitoring, which could have a great impact on patients' quality of life. This is the concept of pervasive monitoring [21]. Such an approach should represent a lower cost for health systems, thanks to the reduced number of visits and hospital stays patients will require. Different scenarios can be envisaged in the continuous search to meet technological challenges through miniaturization, from micro- to nanotechnologies, intelligence, and autonomy [22]. As has been pointed out, special interest has been taken in developments that address chronic illness [48], where traditional approaches are not capable of ruling out sudden death events— particularly in the case of cardiovascular illness [49]. The opportunity of detecting symptoms early in patients who are at risk, of monitoring patients who are following a course of treatment, or of tracking how a disease is progressing offers the possibility of preventing the worst case scenarios. In particular, it is of special interest to track the medical parameters of patients who are not following the traditional clinical observation in hospital. The trick is to monitor patients in their routine daily activities.

Another scenario focuses on elderly patients, who constitute a population at risk; as people live longer, the need for medical resources increases. The possibility of monitoring this substantial proportion of the population, remotely, from their homes, will reduce medical cost [50]. However, BSN monitoring is also opening up new options of interest to hospitals. People who undergo surgery may need to be tracked via intensive monitoring immediately before and after the operation—not just while they are confined to bed. Furthermore, the possibility of monitoring the patient's specific intervention zone after surgery is increasingly becoming a topic of interest [51]. However, the development of such solutions is quite complex.

Different papers have been published recently that focus on applications in different fields. These include, for instance, approaches that present multisensing platforms for patients who could be monitored in their daily lives at home [52]. A variety of sensors are placed in a monitored home laboratory, where different activities are tracked to provide information regarding the motion of the patient; however, this is not limited to an option based on wearable sensors. The authors combine such bodily devices with ambient sensors programmed for specific algorithms that have the capability to identify just one activity at a time. Other approaches have focused on single units that consist of different multisensors [53]. In such approaches a single unit placed on the chest of the patient is able to monitor body motion, activity intensity, and other parameters, such as heart rate, or provide electrocardiography (ECG) data, etc.

BSN solutions are not just conceived as discrete external resources. Specific and highly complex solutions are envisaged in the search for devices that are truly implantable in the body and not just positioned on the body. The power consumption is therefore one of the main issues to consider [54] in a BSN mode. Such systems have been powered—but in external devices, where the approach is different than it is for implantable devices [55], as is presented later. Ideally, the change toward removing the use of batteries would be a priority. Concerns about the role of batteries are also presented in Zheng et al. [56], where computing platforms based on electronic textiles are considered as one of the fields of future development. Batteries have a limited lifetime that may be as long as 10 years in some applications, but the ideal approach would be to remove this element and ensure the reliability and operability of the systems without the need to use an element that must be replaced. This is particularly salient when considering implantable devices. The possibility of ensuring a long-term working life of such devices, without the use of the batteries, can improve the quality of life of patients without the need for any surgery.

Nowadays, the concept of new e-health systems already imposes very low-power consumption restrictions on the electronic instrumentation, processing devices, and communication systems in order to extend the operating life of batteries. If the batteries are removed, then the system must be powered by different kinds of energy sources present in the immediate environment—that is, the body itself or specific solutions that will depend on the placement of the sensors. This situation produces new challenges for engineers. Systems that rely on just one power source could be a problem. If that energy source is not always available, the power could fail. So, different approaches should be considered. In one case, if a battery is still used, that power source could be combined with other possible scavenging sources in such a way that, in terms of the operating protocol of the system, the battery lifetime is extended. This could be achieved by recharging the battery when enough energy is recovered, or the battery could be in an open-load configuration in which the required operating energy is supplied by the scavenger module. If the battery is removed altogether, the system must rely on a combination of different scavenger energy sources.

The types of sensors, their location, and the amount of data that must be processed and transmitted define the power budget that must be considered in order to define the power module.

The approach depicted in Figure 2.1 shows not only the monitoring objective of such a system, but also the concept of a closed loop. In such a system, monitoring the patient is not the only objective; the system also has to be able to generate electrical stimulation. This is a new approach. In some sense the monitoring sensor nodes must take measurements periodically while sensors that also have a treatment functionality can operate either periodically or be event driven, which thereby defines different power consumption scenarios. The typical paradigm example of a sensor node with functionality is the artificial pancreas [57]. Another key example is related to micro nerve stimulation and recording, as in previous work [58], but for which bioamplifier circuits and stimulator output stage circuits are being developed nowadays [59] for regenerative microchannel interfaces.

Many examples are currently being developed in the field of implantable devices. Some of them are presented in what follows, paying special attention to the power involved, power supply, and operating ranges. In the field of pacemakers, Lee et al. [60] present a programmable, implantable microstimulator with wireless telemetry for endocardial stimulation in order to detect and correct cardiac arrhythmias. That option lists the global power consumption as 48 μW, but it relies on a rechargeable battery based on RF coupling. Later, an implantable device that is placed deep in the body is presented. Another interesting case is presented in Langel et al. [61]. There, a transcutaneous implantable device is developed based on an infrared radiation (IR) link just a few centimeters from the emitter. The device is completely autonomous, battery free, and powered through an AC signal operating at the industrial, scientific, and medical (ISM) radio band. It incorporates low-frequency operation at 13.56 MHz, with a global power consumption of 270 μW and an application-specific integrated circuit (ASIC) size of 1.4×1.4 mm^2.

An interesting approach is presented in Liao et al. [62]. There, for the particular case of glucose monitoring, a noninvasive solution is presented based on an active contact lens, whose size is limited to 0.5×0.5 mm^2. In that case, the implant is powered by a rectified 2.4 GHz RF power signal source at a distance of 15 cm with a global power consumption of 3 μW. Some specific examples in the case of neuroscience applications are also introduced. In Chiu et al. [63], a CMOS implementation is presented that operates as a stimulator of the dorsal root ganglion, which has been experimentally probed with rats. A battery-free solution is envisaged with an induced 1 MHz RF signal at a distance of 18 mm, at the standard medical implanted communication system frequency (MICS) of 402 MHz. Another example of an inductively powered neural SoC (system on chip) system is presented in Lee et al. [64]. In a 0.5 μm technology, a 4.9×3.3 mm^2 SoC is designed for a 32-channel wireless integrated neural recording system, with a power dissipation of 5.85 mW.

There are other implantable devices that are more oriented toward biomechanical applications: in particular, the resources needed for monitoring prosthetic implants that are used in human beings to replace joints and bones (Figure 2.3). Such implants are designed to have a duration of 20 years, but degradation can lead to the necessity to replace them, with all the associated inconveniences and risks to the patients. The possibility of integrating electronics to monitor fatigue in the implants has clear advantages in protecting against premature mechanical failures. The amount of wear that implants suffer depends of many variables, but, in particular, on the level of activity and weight of the patient. The typical locations of such implants are knee joints, carpal and tarsal, scapula, and hip-socket joints. Arthroplasty applications with monitoring electronics must ensure genuine autonomous long-term operation,

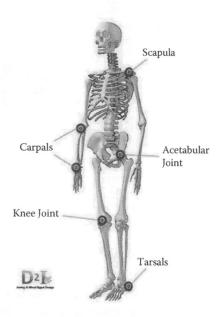

FIGURE 2.3 Implantables for biomechanical applications.

in terms of capturing measurement data that are to be transmitted and powering the system.

Interesting applications have been reported where self-powered systems are based on piezoelectric transducers that use mechanical deformation to sense the strain, but are also a harvesting power source for the system [65]. In such cases, human motion is used to power the electronics and the battery is removed, as is presented later. In Almouahed et al. [66], a knee implant is presented based on the use of piezoceramics able to deliver a maximum average electrical power of 12 mW at the tibial base plate. In Lahuec et al. [67], based on another approach, the estimated power is 1.8 mW. The power levels are low, so the power consumption of the integrated electronics must be very low [68].

The design of an implantable device must contemplate different modules for implementation. The energy levels that are involved in the development of an integrated solution will vary in terms of the final application and the level of intelligence of the device. A system that must work continuously is not the same as a system that works in bursts. Four modules are usually present in an implantable device:

1. The sensor or sensors involved, which fix the type of measurement and the complexity of the front-end instrumentation, which is the signal conditioning
2. The data processing
3. The wireless module
4. The power management unit

If the system has to close the loop, the stimulation electronics—based on DC–DC converters and drivers—are also involved in the power budget. The design of

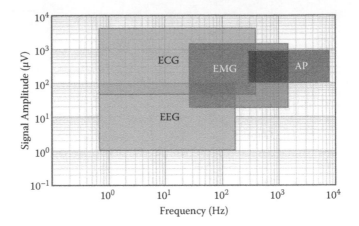

FIGURE 2.4 Some types of medical signals.

the front-end electronics [69] depends on the type of medical signals that must be considered. It can be stated that these signals may vary from a few microvolts to several millivolts, with a frequency band that varies from a few hertz to kilohertz, as depicted in Figure 2.4. Typically, electroencephalogram (EEG) signals ranges from 1 to 50 µV and have low frequencies from 1 to 100 Hz. Electrocardiogram (ECG) signals present the highest voltage level: in the range of millivolts. Electromyogram (EMG) signals and action potentials present the highest frequencies: up to 10 kHz.

The power consumption levels (in terms of energy per operation) depend on the electronics involved. If a microprocessor is needed, then a typical power consumption of 300 µW for commercial solutions, such as the Texas Instruments® MSP430 microprocessor is far from the desired power budget. Favorable examples of specific microprocessors have been presented in Hanson et al. [70] and Zhai et al. [71]. Hanson et al. [70] present the Phoenix Processor, in a 180 nm technology, with a power consumption of 226 nW, but with emphasis on the standby consumption, which is as low as 35.4 pW. It has an area of 915×915 µm² and operates at 0.5 V. It evolved from Zhai et al. [71], where the subthreshold operating region is explored.

Analog–digital converter (ADC) conversion in terms of the needs of sample conversion rates and the submicrowatt transmitters (as a function of the distance between the implanted device and the exterior on the one hand and the data transmission rate on the other) are two of the key elements of these designs. Some pertinent examples show how research aims to find ADCs with lower power consumption. In Song et al. [72], ultra low-power front-end signal conditioning is implemented for an implantable 16-channel neurosensor array, with a maximum power consumption of 50 µW per channel. The ADC is based on a commercial solution derived by Analog Devices®; a 12-bit AD7495 micro-SO8 standard packaged chip. The complete system has a total consumption of 12 mW.

Specific examples of ADCs for implantable applications have been also derived as in Cheong et al. [73]. There, a 400 nW SAR ADC converter, with 8 bits of ENOB (effective number of bits) and 80 kS/s is presented in 180 nm technology.

The implantable blood pressure sensing microsystem developed in Cong et al. [74] achieves 10-bit resolution with an integrated cyclic-ADC converter with 11-bit resolution and a power consumption of 12 µW at 2 V. The pressure sensor has a full-power dissipation of 300 µW. In Trung and Häfliger [75], an ADC converter, based on an integrated ADC, for a blood glucose monitoring implant presents even better performance at just 10.2 nJ per sample, with 10 ENOB, operating at very low frequencies. Another example, in this case in the field of neural signal acquisition, is presented in Muller, Gambini, and Rabaey [76], where the ADC has a power consumption of 240 nW based on a boxcar sampling ADC [77] operating at 20 kS/s.

Regarding the wireless module, different approaches depending on the type of signal captured and the amount of data related to the types of complex signals define a wide range of data transmission constraints from a few samples by second, as in the case of heart rating (typically 25 Sps) or scenarios with data transmission rates of several megabits per second, such as medical imaging [78] or recent developments in endoscopic capsules [79,80].

The review in Bashirullah [81] analyzes in detail the parameterization of the different implantable devices in terms of three key variables: the size of the implantable device, the power involved, and functionality and performance. Two main approaches to communications are considered depending on the type of placement of the implant in the body: inductive coupling [82], when there is short distance between the implant and the exterior of the body, and far field electromagnetic communication [83]. The approach for a subcutaneous implant based on inductive links, where coil misalignment and the effects of the geometry have a great impact on performance [84], is not the same as solutions operating at high frequencies following the MICS protocol [85,86]. For inductive link circuits, the operating frequencies are in the range of a few megahertz, typically in the 13.56 MHz band, following the ISM protocol, with characteristic data modulation methods for transmission, such as BPSK (binary phase-shift keying), ASK (amplitude shift keying), or OOK (on-off keying), and a data transfer rate that is not very fast: from a few kilobits per second (100) to several megabits per second (1.6) [87].

In Simard, Sawan, and Massicotte [88], a high-speed OQPSK is presented with a bit rate of 4.16 Mbps based on an inductive link, with high power levels and multiple coils, operating at 1 MHz for the power link and at the ISM 13.56 MHz frequency for the data. In Jung et al. [89], an implant OOK transmitter is presented that operates at 2.4 GHz in 180 nm technology and is able to transmit at 136 Mbps with a power consumption of 3 mW. As has been pointed out [68], for the particular case of an orthopedic implant, the power consumption for the full electronics must be low. The total power consumption of the full electronics is not stated, but the DC–DC that recovers energy from PZT (lead zirconate titanate) and the ADC converter that presents a quiescent power consumption of 150 nW and an operating consumption of 12.5 µW at 1.8 V for a sample frequency of 4 kHz and an ENOB > 7 bits merit special attention.

2.2.3   POWERING SOLUTIONS FOR HUMAN WEARABLE AND IMPLANTABLE DEVICES

Thus far, we have introduced the concept of energy harvesting and of the basic sources that permit the recovery of energy present in the environment (Figure 2.5).

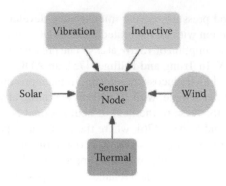

FIGURE 2.5   Combined power sources in a sensor node.

Several applications in the biomedical environment that have been presented have enabled us to visualize the energy needs of these solutions. Specific cases have been discussed, but at this point of development, more specifically, resources that could be used in the strictly biomedical environment. The type of energy source required to power a sensor node varies according to the final application and where the biomedical device is placed. This brings us to three approximations: (a) the possibility of utilizing discrete elements of energy storage, (b) being able to use energy resources present in the environment, and (c) being able to use energy from the human body as a resource. The use of batteries could be a limitation for the envisaged sensor nodes of the future, either on the outside of or implanted within the human body, as in Figure 2.1. The use of large batteries ensures the duration of the system, but the sensor nodes may be too large and heavy. So, smaller solutions with a high enough energy density are needed, combined with ultra low-power electronics solutions that ensure a trade-off between the autonomy of the system and the smart functionality of the sensor, in terms of the sensor, signal processing and communications modules.

Ultra low-power electronics for biomedical applications based on lithium ion batteries are common for nonimplantable solutions, such as BSNs, and also in implantable solutions [90,91]. Lithium ion batteries are divided into two types: (a) single-use batteries, which are placed, for instance, in drug pumps, cardiac defibrillators, and pacemakers; and (b) rechargeable batteries, which are used in artificial hearts. Both types of batteries present an adequate energy density (around 1440 to 3600 J/cm³) and, unlike some other batteries, they do not present the memory effect; that is, they do not need to be discharged completely before a recharge phase. These batteries have a better life cycle than other types, typically with 20,000 discharges and recharge cycles, but with a finite lifetime limitation, which is typically several years for a battery of 1 cm³.

A key aspect for these batteries is the need for battery-management circuitry to ensure the range of operation, as lithium ion batteries are extremely sensitive to overvoltage (maximum 4.2 V) and deep discharge (minimum 2 V), as well as to ensure high energy efficiencies. An application conceived for biomedical applications has been reported [92] with an average power efficiency of 89.7% and a voltage accuracy of 99.9%. Other approaches are under development. Interest is especially focused on fuel cells, such as the methanol fuel cell [93], but these also have their drawbacks. One is the need to replace the external reactant and the oxidant, which is analogous to the problem of recharging batteries. Although higher levels of energy are expected, based on the use of fuels such as methanol with an energy density of 17,600 J/cm^3, the design issues are highly complex and proving to be very expensive.

Supercapacitors are another field that is being explored as an option for biomedical sensors instead of batteries, but they have a low energy density, which is a problem for systems that require a constant power source for long periods of time. In Pandey et al. [94], a sensor is powered via a supercapacitor that is charged wirelessly; in Shanchez, Sodini, and Dawson [95], an energy management ASIC is implemented and tested to manage supercapacitors for implants.

The next issue concerns the power sources that are available in and around the human body. With reference to sources of energy present in the vicinity of the human body that permit the charging of implantable devices, we are able to emphasize recent works, as in Ayazian and Hassibi [108], in which the possibility of using subcutaneous photovoltaic (PV) cells for superficial implants with a recovery capacity of a density to the power of 0.1 $\mu W/mm^2$ in conditions of strong light (sunlight) is discussed. Another interesting approximation is at the level of ultrasound [109], with advantages compared to solutions based on RF.

But we can consider other scenarios. For instance, is it possible to recover energy from human bodily motion? Is it possible to recover energy from the body itself? These are two questions addressed in this section. First, we introduce the concept of a harvester system based on an energy harvester, an energy storage element, and power management regulation (Figure 2.6). Four types of energy harvester based on ambient sources— that is, systems that acquire electrical energy from environmental energy sources—are introduced when this system is described: mechanical, solar, thermal, and RF (the details of which are beyond the scope of the present chapter). The power management unit has two main roles: (a) to generate a raw DC voltage from the harvester unit, essentially acting as a conversion module; and (b) to regulate the output voltage from the energy storage unit to the sensor node electronics. The energy storage unit receives energy from the conversion module and stores it.

Attention is especially focused on body harvesting. The energy can be generated passively or actively. The natural motion of the body, where the conversion is based on mechanical (vibration) energy to electrical energy, is one of the main research topics [24]. In this context, the power that can be harvested when a human being walks or runs has been studied at different locations, with an average of 0.5 mW/cm^3 for hip, chest, elbow, upper arm, and head, and a maximum of 10 mW/cm^3 for the ankle and knee. The mechanical energy can be converted to electrical energy based, for instance, on electromagnetic [24], electrostatic [13], or piezoelectric [15] principles.

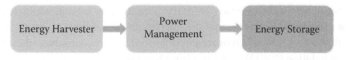

FIGURE 2.6 Schematic for an energy harvesting solution.

TABLE 2.2

Examples of Human Power Actively Generated

Activity	Power Generation
Finger (pushing pen)	0.3 W
Legs (cycling at 25 km/h)	100 W
Hand and arm (Freeplay)	21 W
Hand (AladdinPower)	3.6 W

There are several references to types of design that convert this mechanical energy to electrical energy based on the three methods of transduction just mentioned [96]. The piezoelectric approach is of special interest because high voltages are obtained for low strains with a maximum energy density of 17.7 mJ/cm^3. Several examples are presented in the literature. In particular, in Mhetre et al. [97], piezoelectric microgenerators [98] are envisaged for drug delivery devices or dental applications. Examples of power generated actively are human power through peddling activity [99], or examples such as the Freeplay® or AladdinPower® rechargeable products (see Table 2.2). Walking alone has been analyzed and a prototype implemented [100], but that prototype is excessively large for our purposes.

An extended analysis in terms of the placement of the harvester and the type of everyday activity is presented in Olivares et al. [101]. Based on a commercial cantilever beam harvester from Midé Technologies [102]—the PEH20W Volture—these different scenarios have been analyzed. The greatest level of power extracted (28.74 µW) was from a shank at the instep of the foot while the subject ran fast, and the lowest level of power (0.02 µW) was also from a shank but while performing knee rehabilitation exercises sitting down. As expected, the maximum levels are obtained for activities and placements where higher amplitudes and impacts are produced.

An example of a wearable wireless sensor based on a kinetic energy harvester has been presented [103]. Electromagnetic transduction is used for an average human motion at 0.5 Hz, with a simple architecture implemented via commercial components and a supercapacitor, but not used as a final storage element of the system. It is used to transfer the charge to a smaller supply capacitor, thus improving system start-up. In Zainal Abidin, Hamzah, and Yeop Majlis [104], a MEMS piezoelectric generator is used to harvest energy from vibrations; it also uses supercapacitors as storage elements. An example of a MEMS designed for implantable devices is given in Martínez-Quijada and Chowdhury [105], where it is stated that the microgenerator is able to generate more energy per unit volume than conventional batteries—that is, an RMS power of 390 µW for 1 mm^2 of footprint area and a thickness of 500 µm, which is smaller than the volume of a typical battery in a pacemaker.

The other main source of power from the human body is heat, which is limited by the Carnot efficiency that states that the maximum power that can be recovered is in the range of 2.4 to 4.8 W when other possible sources are available, like arm motion (0.33 W), exhalation (0.40 W), or breathing band (0.42 W). In the case of body heat, thermoelectric generators are used [106]. Specific designs must be considered when working with thermoelectric generators in these conditions, as the voltages generated are very small and DC–DC converters with a specific step-up are conceived in order to boost the voltage [107]. An example of a thermogenerator is Thermolife®, which generates 60 µW/cm² for a temperature difference of 5°C [11]. It should also be stated that the recovery depends on the specific placement of the thermoelectric generator. Placing it in the neck is not the same as placing it in the head, both of which are parts of the body that are warmer than other zones. The recovery range is 200 to 320 mW for the neck and 600 to 960 mW for the head (three times the surface area of the neck).

There are other approximations in the area of the recovery of energy from the human body itself. From the concept of a fuel cell, the biogenerator for implantable devices has emerged. In some ways, the basic concept is the use of fluids in the body as a fuel source for the fuel cell; this would be an inexhaustible energy source. An interesting approach is the use of glucose as a fuel source and the oxygen dissolved in blood [110,111]. Advanced approaches also explore a shift to the use of white blood cell capacities in biofuel cells [112] or approaches such as that in Muller, Gambini, and Rabaey [76], where the fuel cell is based on the use of a microorganism to convert the chemical energy of glucose into electrical energy in a PDMS (polydimethylsiloxane) (plasma desorption ionization mass spectrometry) structure [113].

2.3 MULTISOURCE SELF-POWERED DEVICE CONCEPTION

At this point, a sensor node could be powered by following different approaches; the choice depends on the placement of the sensor, which defines its accessibility, its size, and its weight. A battery, a supercapacitor, and a fuel cell are possible choices on the one hand. On the other, the use of some kind of energy harvesting source that recovers energy from the body, mostly based on vibration to electrical conversion or thermoelectric generation, is also envisaged. However, there are some other approaches that can also be considered, depending on the placement of the sensor; these include, the possibility of using environmental resources to harvest energy from for the sensor node, such as light [9,114] or radio waves [10].

The approach introduced previously is RF coupling, both as a way to supply energy to the sensor node and to use the RF link for communications purposes, typically for short distances between the master and slave modules and typically in the 13.56 MHz ISM band. For instance, at 4 MHz and a distance of 25 mm, for subcutaneous powering of a biomedical device, the energy recovered was 5 mW [115]. A particular example of this is introduced in Zhang et al. [116]. There, the rectifier module is designed to work at the 915 MHz ISM bandwidth (only in region 2). The performance of the RF source is quite small, from a typical 4 µW/cm² for GSM to 1 µW/cm² for the WiFi band. For coils, typical values are lower than 1 µW/cm², but with as much as 1 mW for close inductive coils (a few centimeters). In the 915 MHz ISM bandwidth, at 1.1 m, the energy recovered is around 20 µW [12]. New approaches are being developed. In particular, the use of ultrasonic powering instead of RF powering is of great interest. In Zhu, Moheimani, and Yuce [117] and Yong, Moheimani, and Yuce [109], 1 V is generated with a power capability of 21.4 nW.

At this point the issue that we need to consider is the possibility of placing or using more than just one energy harvester. This approach, combining indoor solar cells, RF coupling at the ISM 13.56 MHz band, and mechanical vibration based on piezoelectric transducer, was presented in Colomer-Farrarons et al. [1], together with experimental results. The system does not rely on only one harvesting source, and the objective is to use different energy harvesters. A specific power management unit has been designed for a multisource and multiload energy harvesting system with the aim of establishing a minimum number of conversion stages and magnetic components, and of developing a specific algorithm [118]. An approach to the envisaged platform architecture for a multiharvesting WSN has been presented [119]. Another interesting example, commented on in Lahuec et al. [67], has the particularity that it combines the possibility of recovering energy from the motion of a human being based on PZTs, with recovering energy from the RF signal, depending on the operating mode; however, no more details have been reported.

The combination of a vibration energy harvester and charging circuitry for a lithium ion battery has been presented [120]. The architecture proposed in references 1 and 121 has been used to demonstrate the feasibility of combining different harvester sources for a 1 cm^3 multisensor node. This is a generic architecture envisaged for a generic self-powered smart sensor that could be applied in different fields of applications, such as the integrity of structures, surveillance, and medical devices and implants. Electromagnetic coupling, with an indoor solar-cell module and a piezoelectric generator, was used. The platform envisaged is defined as a multiharvesting power chip (MHPC) with a total power consumption of 160 µW. The integrated circuit is designed with 0.13 µm technology, which is a low-voltage technology up to 3.3 V. The capacity of a small system to recover a few microwatts from the energy present in the environment of the multisensor node, combining vibration, light, and RF, is a proof of concept and recovers a total power of more than 1 mW.

The system is initially analyzed with each different source working alone. The system recovers 360 µW from the piezoelectric generator, which is a commercial PZT (QP40W) [122], operating at 7 m/s^2 at 80 Hz. Then, two indoor solar cells [123] under indoor conditions of 1500 lx are used, with a total harvesting power of 2.76 mW. Finally, an RF link generator (TRF7960) emitting at full power (200 mW) with a distance between the base station and the antenna receiver of 25 mm is used and recovers 4.5 mW. The total power expected to be recovered from the combination of the three harvested sources would be 7.8 mW, but the experimentally recovered power was 6.2 mW [1,121]. The difference of 1.6 mW between the theoretical total and the actual amount of power that was recovered is due to variations in the effective load conditions for the light and RF modules.

This suggests that research is needed into the trade-off between the implementation of peak power tracking circuitry for the light and RF modules and the improvement that could be achieved, compared with the cost in silicon area and the power consumption of each module. The three main elements in the design of the MHCP ASIC are as follows:

1. A bandgap reference circuit is based on peak regulation.
2. A linear dropout (LDO) regulator i based on the bandgap reference circuit, which uses a PMOS switch and an error amplifier [1]. The PMOS (p-type metal–oxide semiconductor) switch of the LDO regulator is placed at the

output stage of the simple two-stage amplifier, where MN1 and MN2 define the ground current. These elements have a power requirement of 30 μW.

3. Two integrated rectifiers must be placed in the MHCP chip, one for AC/DC rectification of the piezoelectric generator signal and the other one for the RF-coupled signal.

The MHCP has its own control unit that plays the role of the power management unit, based on the concept presented in Colomer-Farrarons et al. [4]. The system combines a control switch for each of the energy harvester channels in order to be able to operate as a single and independent harvesting channel, with independent storage elements for each channel—that is, multiple storage device configuration (MSD mode of operation) [1] or combining of all them into a single storage device (SSD mode of operation). In SSD mode, the PMU controls the energy stored in the single storage element (a supercapacitor) and transfers energy when an adequate voltage is reached in the capacitor. When the voltage reaches the low-value threshold (Vmin), the system opens the charge transfer to the load until the maximum voltage level (Vmax) is reached again. The PMU also incorporates power on reset (POR). This circuit is used to reduce the power consumption of the module during the start-up phase.

Based on this MHCP ASIC conception, a CMOS architecture for an implantable device [124] is envisaged. Nowadays, interest in nanobiosensors is increasing in the field of medical diagnosis. The development of such devices and the telemedicine environments that can be derived from them has great market potential, as has been pointed out before. Different approaches are required for discrete, small cubic centimeter devices and for implantable devices; the performance, communications capabilities, etc. are very different. The size of the implantable device is envisaged as that of a capsule, ideally less than 4.5 cm long and 2.5 cm in diameter, following the same philosophy as some subcutaneous implantable contraceptive devices such as Norplant®, Jadelle®, and Implanon®.

One proposal is an event-detector implantable system, or an event detector that works as an alarm. When the concentration analyzed moves out of the range of accepted values and reaches a threshold value, the alarm is activated. The proposed generic implantable architecture is presented in Colomer-Farrarons and Miribel-Catala [121]. It is composed of three BioSensor electrodes, an antenna, and the electronic modules. Such a system combines different modules. The antenna and the AC/DC module are used to supply energy to the device (inductive powering) and the communication setup (backscattering) based on AM modulation. Then, a low-voltage and low-power potentiostat is integrated. The BioSensor is the only part of the implantable device interacting with the biological environment. It detects the desired target generating an electrical signal. The BioSensor's design must be carefully selected as a function of the target sample to be detected and the total size of the device.

Several BioSensor configurations formed by simply two, three, or four electrodes can be used for single target detection, and more complex array structures of microsensors can be introduced for multianalyte detection. The antenna and its associated electronics are used for two main aspects. Firstly, it is used to supply energy to the implantable device (inductive powering) working together with an integrated AC/DC module. Secondly, the same antenna transmits the information to the external reader through the communications electronic setup (backscattering). In this scenario, one antenna is used for both power generation and communication. This reduces the antenna operation frequency to tens of millihertz due to the inductive power drop caused by the human skin. Moreover, the amount of transmitted information is limited and the size of the antenna is considerable.

Another scenario considers the use of two antennas in the same implantable device: one for the communication and the other for powering. In this case, the communication link can be established around hundreds of megahertz (usually in the 400 MHz ISM band), allowing higher communication rates and reducing the size of the antenna. On the other side, the second antenna is focused to power the electronics through a dedicated inductive link operating at lower frequencies than the communication antenna. In that way, each antenna can be optimized for its functionality. Then, the integrated electronics is introduced to drive the BioSensor and to generate the data to be transmitted. Usually, a low-voltage, low-power potentiostat circuit or similar instrumentations are used to control the implantable sensors.

2.4 SUMMARY AND CONCLUSIONS

This chapter introduces a review in the state of the art of energy harvesting and focuses interest on the idea of using different types of energy harvesting sources to drive ultra low-power electronics. This approach is followed in order to develop new portable biodevices for human wearable and implantable devices. In that way, it would be possible to envisage new autonomous smart healthcare portable instruments able to be adapted easily to daily human activity. The importance of harvest energy from normal human activities introduces the enormous advantage of reducing bulky batteries from small handheld devices, allowing miniaturization and weigh reduction of the portable instruments.

REFERENCES

1. Colomer-Farrarons, J., Miribel-Catala, P., Saiz-Vela, A., and Samitier, J. 2011. A multiharvested self-powered system in a low-voltage low-power technology. *IEEE Transactions on Industrial Electronics* 58 (9): 4250–4263.
2. Sazonov, E. 2009. Self-powered sensors for monitoring on highway bridges. *IEEE Sensors Journal* 9 (11): 1422–1429.
3. Wang, W. et al. 2010. Autonomous wireless sensor network based building energy and environment monitoring system design (paper presented at the International Conference on Environmental Science and Information Application Technology, Wuhan, China, July 17–18).
4. Colomer-Farrarons, J. et al. 2008. Power-conditioning circuitry for a self powered system based on micro PZT generators in a 0.13 µm low-voltage low-power technology. *IEEE Transactions on Industrial Electronics* 55 (9): 3249–3257.
5. Yang, G.-Z. 2006. *Body sensor networks*. London: Springer.
6. Dey, A. K., and D. Estrin. 2011. Perspectives on pervasive health from some of the field's leading researchers. *IEEE Pervasive Computing* 10 (2): 4–7.

7. Mehraeen, S., S. Jagannathan, and K. A. Corzine. 2010. Energy harvesting from vibration with alternate scavenging circuitry and tapered cantilever beam. *IEEE Transactions on Industrial Electronics* 57 (3): 820–830.
8. E. Carlson, K. Strunz, and B. Otis. 2009. 20 mV Input boost converter for thermoelectric energy harvesting (paper presented in the Symposium on VLSI Circuits, Seattle, WA, June 16–18).
9. Nasiri, A., S. A. Zabalawi, and G. Mandic. 2009. Indoor power harvesting using photovoltaic cells for low-power applications. *IEEE Transactions on Industrial Electronics* 56 (11): 4502–4509.
10. Sogorb, T., J. V. Llario, J. Pelegri, R. Lajara, and J. Alberola. 2008. Studying the feasibility of energy harvesting from broadcast RF station for WSN (paper presented in the IEEE Instrumentation and Measurement Technology Conference, Victoria, BC, Canada, May 12–15).
11. Roundy, S., D. Steingart, L. Frechette, P. Wright, and J. Rabaey. 2004. Power sources for wireless sensors networks (paper presented in the First European Workshop on Wireless Sensors Networks, Berlin, Germany, January 19–21).
12. Yeatman, E. M. 2007. Energy scavenging for wireless sensor nodes (paper presented in the 2nd International Workshop on Advances in Sensors and Interface, Bari, Italy, June 17–26).
13. Kiziroglou, M. E., C. He, and E. M. Yeatman. 2009. Rolling rod electrostatic microgenerator. *IEEE Transactions on Industrial Electronics* 56 (4): 1101–1108.
14. Le, T. T. et al. Piezoelectric micro-power generation interface circuits. *IEEE Journal of Solid-State Circuits* 41 (6): 1411–1420.
15. Garbuio, L. et al. 2009. Mechanical energy harvester with ultralow threshold rectification based on SSHI nonlinear technique. *IEEE Transactions on Industrial Electronics* 56 (4): 1048–1056.
16. Carrara, S. et al. 2012. Developing highly integrated subcutaneous biochips for remote monitoring of human metabolism (paper presented in the IEEE Sensors, Taipei, Taiwan, October 28–31).
17. Curty, J. P., N. Noehl, C. Dehollain, and M. Declercp. 2005. Remotely powered addressable UHF RFID integrated system. *IEEE Journal of Solid-State Circuits* 40 (11): 2193–2202.
18. Thermo Life LPTG. Thermo Life Energy Corp. http://www.poweredbythermolife.com/
19. Van Hoof, C., V. Leonov, and R. J. M. Vullers. 2009. Thermoelectric and hybrid generators in wearable devices and clothes (paper presented in the Sixth International Workshop on Wearable and Implantable Body Sensor Networks, Berkeley, CA, June 3–5).
20. Nasiri, A., S. A. Zabalawi, and G. Mandic. 2009. Indoor power harvesting using photovoltaic cells for low-power applications. *IEEE Transactions on Industrial Electronics* 56 (11): 4502–4509.
21. Garcia-Morchon, O., T. Falck, T. Heer, and K. Wehrle. 2009. Security for pervasive medical sensor networks (paper presented in the International Mobile and Ubiquitous Systems Conference: Networking & Services, MobiQuitous, Toronto, Canada, July 13–16).
22. Penders, J. et al. 2008. Human++: From technology to emerging health monitoring concepts (paper presented in the International Summer School and Symposium on Medical Devices and Biosensors, Hong Kong, China, June 1–3).
23. Mhetre, M. R., N. S. Nagdeo, and H. K. Abhyankar. 2011. Micro energy harvesting for biomedical applications: A review (paper presented in the IEEE International Conference on Electronics Computer Technology, Kanyakumari, India, April 8–10).

24. Christmann, J. F., E. Beigné, C. Condemine, and J. Willemin. 2010. An innovative and efficient energy harvesting platform architecture for autonomous microsystems (paper presented in the IEEE International NEWCAS Conference, Montreal, Canada, June 20–23).

25. Lu, C., V. Raghunathan, and K. Roy. 2011. Efficient design of micro-scale energy harvesting systems. *IEEE Journal on Emerging and Selected Topics in Circuits and Systems* 1 (3): 254–266.

26. Jornet, J. M. 2012. A joint energy harvesting and consumption model for self-powered nano-devices in nanonetworks (paper presented in the International Conference on Communications, Ottawa, Canada, June 10–15).

27. Khoshnoud, F., and C. W. de Silva. 2012. Recent advances in MEMC sensor technology-mechanical applications. *IEEE Instrumentation & Measurement Magazine* 15 (2): 14–24.

28. Benecke, S., J. Ruckschloss, N. F. Nissen, and K.-D. Lang. 2012. Energy harvesting on the way to a reliable and green micro energy source (paper presented in the Electronics Goes Green Conference, Berlin, Germany, September 9–12).

29. Lu, J., Y. Zhang, T. Itoh, and R. Maeda. 2011. Design, fabrication and integration of piezoelectric MEMS devices for applications in wireless sensor network (paper presented in the International Symposium on Design, Test, Integration and packaging of MEMS/MOEMS, Aix-en-Provence, France, May 1–13).

30. Trolier-McKinstry, S. et al. 2011. Designing piezoelectric films for micro electro-mechanical systems. *IEEE Transactions on Ultrasonics, Ferroelectrics, and Frequency Control* 58 (9): 1782–1792.

31. Liu, S. W., S. W. Lye, and J. M. Miao. 2012. Sandwich structured electrostatic/electrets parallel-plate power generator for low acceleration and low frequency vibration energy harvesting (paper presented in the International Conference on Electro Mechanical Systems, Cancun, Mexico, June 23–27).

32. Peterson, K., and G. A. Rincon-Mora. 2012. High-damping energy-harvesting electrostatic CMOS charger (paper presented in the IEEE International Symposium on Circuits and Systems, COEX, Seoul, Korea, May 20–23).

33. Martorell, J. 2012. A photonic nano-structuring approach to increase energy harvesting for organic photovoltaic cells (paper presented in the International Conference on Transparent Optical Networks, Coventry, England, July 2–5).

34. Lu, C., S. P. Park, V. Raghunathan, and K. Roy. 2012. Low-overhead maximum power point tracking for micro-scale solar energy harvesting systems (paper presented in the International Conference on VLSi Design, Hyderabad, India, January 7–11).

35. Chen, B. 2012. Nanomaterials for green energy: Next generation energy conversion and storage. *IEEE Nanotechnology Magazine* 9: 4–7.

36. Bareiss, M. et al. 2011. Energy harvesting using nano antenna array (paper presented in the IEEE International Conference on Nanotechnology, Portland, OR, August 15–19).

37. Paul, D. J. et al. 2012. Si/SiGe nanoscale engineered thermoelectric materials for energy harvesting (paper presented in the IEEE International Conference on Nanotechnology, Birmingham, U.K., August 20–23).

38. AbdeElFattah, M., A. Mohieldin, A. Emira, and E. Sanchez-Sinencio. 2011. A low-voltage charge pump for micro scale thermal energy harvesting (paper presented in the IEEE International Symposium on Industrial Electronics, Gdansk, Poland, June 27–30).

39. Zargham, M., and P. G. Gulak. 2012. High-efficiency CMOS rectifier for fully integrated mW wireless power transfer (paper presented in the IEEE International Symposium on Circuits and Systems, COEX, Seoul, Korea, May 20–23).

40. Maurath, D., P. F. Becker, D. Spreemann, and V. Manoli. 2012. Efficient energy harvesting with electromagnetic energy transducers using active low-voltage rectification and maximum power point tracking. *IEEE Journal of Solid-State Circuits* 47 (6): 1369–1380.

41. Leicht, J., D. Maurath, and V. Manoli. 2012. Autonomous and self-starting efficient micro energy harvesting interface with adaptive MPPT, buffer monitoring, and voltage stabilization (paper presented in the ESSCIRC Conference, Bordeaux, France, September 17–21).

42. Le, H., N. Fong, and H. C. Luong. 2011. An energy harvesting circuit for GHz on-chip antenna measurement (paper presented in the IEEE International Symposium on Radio-Frequency Integration Technology, Beijing, China, November 30–December 2).

43. Frizzell-Makowski, L. J., R. A. Shelsby, J. Mann, and D. Scheidt. 2011. An autonomous energy harvesting station-keeping vehicle for persistent ocean surveillance (paper presented in the OCEANS Conference, Hawaii, September 19–22).

44. Raisigel, H. et al. 2010. Autonomous wireless sensor node for building climate conditioning application (paper presented in the International Conference on Sensor Technologies and Applications, Venice/Mestre, Italy, July 18–25).

45. vom Boegel, G., F. Meyer, and M. Kemmerling. 2012. Batteryless sensors in building automation by use of wireless energy harvesting (paper presented in the IEEE International Symposium on Wireless Systems, Offenburg, Germany, September 20–21).

46. Zhu, D. et al. 2013. Novel miniature airflow energy harvester for wireless sensing applications in buildings. *IEEE Sensors Journal* 13 (2): 691–700.

47. Lee, Y. et al. 2013. A modular 1 mm³ die-stacked sensing platform with low power I²C inter-die communication and multi-modal energy harvesting. *IEEE Journal of Solid-State Circuits* 48 (1): 229–243.

48. Koutkias, V. G. et al. 2010. A personalized framework for medication treatment management in chronic care. *IEEE Transactions on Information Technology in Biomedicine* 14 (2): 464–472.

49. Zhou, H.-Y., and K.-M Hou. 2010. Pervasive cardiac monitoring system for remote continuous heart care (paper presented in the International Conference on Bioinformatics and Biomedical Engineering, Chengdu, China, June 18–20).

50. Zwifel, P., S. Felder, and M. Meiers. 1999. Ageing of population and health care expenditure: A red herring? *Health Economics* 8: 485–496.

51. Allen, M. G. 2005. Micromachined endovascularly implantable wireless aneurysm pressure sensors: From concept to clinic (paper presented in the 13th International Conference on Solid-State Sensors, Actuators and Microsystems, Transducers, Seoul, Korea, June 5–9).

52. El Helw, M. et al. 2009. An integrated multi-sensing framework for pervasive healthcare monitoring (paper presented in the International Conference on Pervasive Computing Technologies for Healthcare, London, UK, April 1–3).

53. Chuo, Y. et al. 2010. Mechanically flexible wireless multisensor platform for human physical activity and vitals monitoring. *IEEE Transactions on Biomedical Circuits and Systems* 4 (5): 281–294.

54. Mandal, S., L. Turicchia, and R. A. Sarpeshkar. 2010. A low-power, battery-free tag for body sensor networks. *IEEE Pervasive Computing* 9 (1): 71–77.

55. Fang, Q. et al. 2011. Developing a wireless implantable body sensor network in MICS band. *IEEE Transactions on Information Technology in Biomedicine* 15 (4): 567–576.

56. Zheng, N. et al. 2010. Enhancing battery efficiency for pervasive health-monitoring systems based on electronic textiles. *IEEE Transactions on Information Technology in Biomedicine* 14 (2): 350–359.

57. Ricotti, L. et al. 2011. A novel strategy for long-term implantable pancreas (presented at the International IEEE Conference Engineering in Medicine and Biology Society).
58. Kovacs, G. T. A., C. W. Storment, and J. M. Rosen. 1992. Regeneration microelectrode array for peripheral nerve recording and stimulation. *IEEE Transactions on Biomedical Engineering* 39 (9): 893–902.
59. Gerald, F. et al. 2012. A regenerative microchannel neural interface for recording from and stimulating peripheral axons in vivo. *Journal of Neural Engineering* 9 (1): 016010.
60. Lee, S.-Y. et al. 2011. A programmable implantable micro-stimulator SoC with wireless telemetry: Application in closed-loop endocardial stimulation for cardiac pacemaker (paper presented in the International Solid-State Circuits Conference ISSCC, San Francisco, CA, February).
61. Langel, S. et al. 2011. An AC-powered optical receiver consuming 270 µw for transcutaneous 2 Mb/s data transfer (paper presented in the International Solid-State Circuits Conference ISSCC, San Francisco, CA, February).
62. Liao, Y-T., H. Yao, B. Parviz, and B. Otis. 2010. A 3 µW wirelessly powered CMOS glucose sensor for an active contact lens (paper presented in the International Solid-State Circuits Conference ISSCC, San Francisco, CA, February).
63. Chiu, H.-W. et al. 2010. Pain control on demand based on pulsed radio-frequency stimulation of the dorsal root ganglion using a batteryless implantable CMOS SoC. *IEEE Transactions on Biomedical Circuits and Systems* 4 (6): 350–359.
64. Lee, S. B. et al. 2010. An inductively powered scalable 32-channel wireless neural recording system-on-a-chip for neuroscience applications. *IEEE Transactions on Biomedical Circuits and Systems* 4 (6): 360–371.
65. Chao, P. C. 2011. Energy harvesting electronics for vibratory devices in self-powered sensors. *IEEE Sensors Journal* 11 (12): 3106–3121.
66. Almouahed, S. et al. 2011. The use of piezoceramics as electrical energy harvesters within instrumented knee implant during walking. *IEEE/ASME Transactions on Mechatronics* 16 (5): 799–807.
67. Lahuec, C. et al. 2011. A self-powered telemetry system to estimate the postoperative instability of a knee implant. *IEEE Transactions on Biomedical Engineering* 58 (3): 822–825.
68. Chen, H. et al. 2009. Low-power circuits for the bidirectional wireless monitoring system of the orthopedic implants. *IEEE Transactions on Biomedical Circuits and Systems* 3 (6): 437–443.
69. Yazicioglu, R. F., C. Van-Hoof, and R. Puers. 2009. *Biopotential readout circuits for portable acquisition systems,* Analog Circuits and Signal Processing Series. New York: Springer.
70. Hanson, S. et al. 2009. A low-voltage processor for sensing applications with picowatt standby mode. *IEEE Journal of Solid-State Circuits* 4 (4): 1145–1155.
71. Zhai, B. et al. 2009. Energy-efficient subthreshold processor design. *IEEE Transactions on Very Large Scale Integration (VLSI) Systems* 17 (8): 1127–1137.
72. Song, Y.-K. et al. 2009. Active microelectronic neurosensor arrays for implantable brain communication interfaces. *IEEE Transactions on Neural Systems and Rehabilitation Engineering* 17 (4): 339–345.
73. Cheong, J.-H. et al. A 400 nW 19.5 fJ/Conversion-step 8-ENOB 80 kS/s SAR ADC in 0.18 µm CMOS. *IEEE Transactions on Circuits and Systems II: Express Briefs* 58 (7): 407–411.
74. Cong, P., N. Chaimanonart, W. H. Ko, and D. J. Young. 2009. A wireless and batteryless 10-bit implantable blood pressure sensing microsystem with adaptive RF powering for real-time laboratory mice monitoring. *IEEE Transactions of Solid-State Circuits* 44 (12): 3631–3644.

75. Trung, N. T., and P. Häfliger. 2011. Time domain ADC for blood glucose implant. *Electronics Letters* 47 (26): S18–S20.
76. Muller, R., S. Gambini, and J. M. Rabaey. 2012. A 0.013 mm², 5 µW, DC-coupled neural signal acquisition IC with 0.5 V supply. *IEEE Journal of Solid-State Circuits* 47 (1): 232–243.
77. Ezekwe, C. D., and E. B. Boser. 2008. A mode-matching ΣΔ closed-loop vibratory gyroscope readout interface with a 0.004 deg/s/rtHz noise floor over a 50 Hz band. *IEEE Journal of Solid-State Circuits* 43 (12): 3039–3048.
78. Liu, Y.-H. 2008. A 3.5 mW 15 Mbps O-QPSK transmitter for real-time wireless medical imaging applications (paper presented in the IEEE Custom Integrated Circuits Conference, San Jose, CA, September 21–24).
79. Diao, S. et al. 2011. A low-power, high data-rate CMOS ASK transmitter for wireless capsule endoscopy (paper presented in the *Defense Science Research Conference and Expo, DSR*, pp.1–4, Singapore, August).
80. Wang, Q., K. Wolf, and D. Plettermeier. 2010. An UWB capsule endoscope antenna design for biomedical communications (paper presented in the International Symposium on Applied Sciences in Biomedical and Communication Technologies, ISABEL, Rome, Italy, November 2010).
81. Bashirullah, R. 2010. Wireless implants. *IEEE Microwave Magazine* 2010:S14–S23.
82. Lenaerts, B., and R. Puers. 2009. *Omnidirectional inductive powering for biomedical implants,* Analog Circuits and Signal Processing Series. New York: Springer.
83. Alomainy, A., and Y. Hao. 2009. Modeling and characterization of biotelemetric radio channel from ingested implants considering organ contents. *IEEE Transactions on Antennas and Propagation* 57 (4): 999–1005.
84. Fotopoulou, K., and B. W. Flynn. 2011. Wireless power transfer in loosely coupled links: Coil misalignment model. *IEEE Transactions on Magnetics* 47 (2): 416–430.
85. Walk, J. et al. 2011. Remote powering systems of medical implants for maintenance free healthcare applications (paper presented in the 41st European Microwave Conference, Manchester, UK, October 10–13).
86. Abouei, J. et al. 2011. Energy efficiency and reliability in wireless biomedical implant systems. *IEEE Transactions on Information Technology in Biomedicine* 15 (3): 456–466.
87. Sawan, M., and B. Gosselin. 2008. CMOS circuits for biomedical implantable devices. In *VLSI circuits for biomedical applications,* 45–74. Norwood, MA: Artecht House, Inc.
88. Simard, G., M. Sawan, and D. Massicotte. 2010. High-speed OQPSK and efficient power transfer through inductive link for biomedical implants. *IEEE Transactions on Biomedical Circuits and Systems* 4 (3): 192–200.
89. Jung, J. et al. 2010. 22 pJ/bit Energy-efficient 2.4 GHz implantable OOK transmitter for wireless biotelemetry systems: In vitro experiments using rat skin-mimic. *IEEE Transactions on Microwave Theory and Techniques* 58 (12): 4102–4111.
90. Spillman, D. M., and E. S. Takeuhi. 1999. Lithium ion batteries for medical devices (paper presented in the Fourteenth Annual Battery Conference on Applications and Advances, Long Beach, CA, January 12–15).
91. Rubino, R. S., H. Gan, and E. S.Takeuchi. 2002. Implantable medical applications of lithium-ion technology (paper presented in the Seventeenth Annual Battery Conference on Applications and Advances, Long Beach, CA, January 18).
92. Valle, D. Do., C. T. Wentz, and R. Sarpeshkar. 2011. An area and power-efficient analog li-ion battery charger circuit. *IEEE Transactions on Biomedical Circuits and Systems* 5 (2): 131–137.
93. Yuming, Y. et al. 2011. Low-power fuel delivery with concentration regulation for micro direct methanol fuel cell. *IEEE Transactions on Industry Applications* 47 (3): 1470–1479.

94. Pandey, A., F. Allos, A. Patrick Hu, and D. Budgett. 2011. Integration of supercapacitors into wirelessly charged biomedical sensors (paper presented in the 6th IEEE Conference on Industrial Electronics and Applications, lloc, data).
95. Shanchez, W., C. Sodini, and J. L. Dawson. 2010. An energy management IC for bio-implants using ultracapacitors for energy storage (paper presented in the IEEE Symposium on VLSI Circuits, Honolulu, HI, June 16–18).
96. Roundy, S., P. K. Wright, and K. S. J. Pister. 2002. Micro-electrostatic vibration-to-electricity converters (paper presented in the International Mechanical Engineering Congress & Exposition, New Orleans, LA, November 17–22).
97. Mhetre, M. R. et al. 2011. Micro energy harvesting for biomedical applications: A review (paper presented in the International Conference on Electronics Computer Technology, lloc, data).
98. Zhang, J. Y. et al. 2012. Microstructure and piezoelectric properties of AIN thin films grown on stainless steel for the application of vibration energy harvesting. *IET Micro & Nano Letters* 7 (12): 1170–1172.
99. Kazazian, T., and A. J. Jansen. 2004. Eco-design and human-powered products (paper presented in the Electronics Goes Green Conference, Berlin, Germany, September 6–10).
100. Donelan, J. M., V. Naing, and Q. Li. 2009. Biomechanical energy harvesting (paper presented in the IEEE Radio and Wireless Symposium, San Diego, CA, January 18–22).
101. Olivares, A. et al. 2010. A study of vibration-based energy harvesting in activities of daily living (paper presented in the Pervasive Computing Technologies for Healthcare, Munich, Germany, March).
102. htlp://www.mide.comlproducts/volture/peh20w.php
103. Olivo, J., D. Brunelli, and L. Benini. 2010. A kinetic energy harvester with fast start-up for wearable body-monitoring sensors (paper presented in the 4th International Conference on Pervasive Computing Technologies for Healthcare, Lausanne, Switzerland, March 22–25).
104. Zainal Abidin, H. E., A. A. Hamzah, and B. Yeop Majlis. 2011. Design of interdigitaed structures supercapacitor for powering biomedical devices (paper presented in the IEEE Regional Symposium on Micro and Nanoelectronics, Kota Kinabalu, Malaysia, September 28–30).
105. Martínez-Quijada, J., and S. Chowdhury. 2007. Body-motion driven MEMS generator for implantable biomedical devices (paper presented in the Canadian Conference on Electrical and Computer Engineering, lloc, data).
106. Lay-Ekuakille, A. et al. 2009. Thermoelectric generator design based on power from body heat for biomedical autonomous devices (paper presented in the IEEE International workshop on Medical Measurements and Applications, lloc, data).
107. Ramadass, Y. K., and A. P. Chandrakasan. 2011. A battery-less thermoelectric energy harvesting interface circuit with 35 mV startup voltage. *IEEE Journal of Solid-State Circuits* 46 (1): 333–341.
108. Ayazian, S., and A. Hassibi. 2011. Delivering optical power to subcutaneous implanted devices (paper presented in the IEEE International Conference Engineering in Medicine and Biology Society, lloc, data).
109. Yong Shu, R., S. O. Moheimani, and M. Rasit Yuce. 2011. A 2-DOF MEMS ultrasonic energy harvester. *IEEE Sensors Journal* 11 (1): 155–161.
110. Ravariu, C., C. Ionescu-Tirgoviste, and F. Ravariu. 2009. Glucose biofuels properties in the bloodstream in conjunction with the beta cell electro-physiology (paper presented in the International Conference on Clean Electrical Power, Boston, MA, August 20–September 3).
111. Stetten, F. V. et al. 2006. A one-compartment, direct glucose fuel cell for powering long-term medical implants (paper presented in the 19th International Conference on Micro Electro Mechanical Systems, MEMS, Istambul, Turkey, January 22–26).

112. Justin, G. A., Y. Zhang, M. Sun, and R. Sclabassi. 2005. An investigation of the ability of white blood cells to generate electricity in biofuel cells (paper presented in the IEEE 31st Annual Northeast Bioengineering Conference, Hoboken, NJ, April 2–3).
113. Siu, C. P. B., and M. Chiao. 2008. A microfabricated PDMS microbial fuel cell. *Journal of Microelectromechanical Systems* 17 (6): 1329–1341.
114. Bertacchini, A., D. Dondi, L. Larcher, and P. Pavan. 2008. Performance analysis of solar energy harvesting circuits for autonomous sensors (paper presented in the 34th Annual Conference of IEEE Industrial Electronics, Orlando, FL, November 10–13).
115. Sauer, C., M. Stanacevic, G. Cauwenberghs, and N. Thakor. 2005. Power harvesting and telemetry in CMOS for implant devices. *IEEE Transactions on Circuits and Systems* 52 (12): 2605–2613.
116. Zhang, X. et al. 2010. An energy-efficient ASIC for wireless body sensor networks in medical applications. *IEEE Transactions on Biomedical Circuits and Systems* 4 (1): 11–18.
117. Zhu, Y., S. O. R. Moheimani, and M. R. Yuce. 2010. Ultrasonic energy transmission and conversion using a 2-D MEMS resonator. *IEEE Device Letters* 31 (4): 374–376.
118. Saggini, S., and P. Mattavelli. 2009. Power management in multi-source multi-load energy harvesting systems (paper presented in the European Conference on Power Electronics and Applications, Barcelona, Spain, September 8–10).
119. Christmann, J. F., E. Beigné, C. Condemine, and J. Willemin. 2010. An innovative and efficient energy harvesting platform architecture for autonomous microsystems (paper presented in the International NEWCAS Conference, Montréal, Canada, June 20–23).
120. Torres, E. O, and G. A. Rincon-Mora. 2009. Electrostatic energy-harvesting and battery-charging CMOS system prototype. *IEEE Transactions on Circuits and Systems I: Regular Papers* 56 (9): 1938–1948.
121. Colomer-Farrarons, J., and P. Miribel-Catala. 2011. *A CMOS self-powered front-end architectures for subcutaneous event-detector devices: Three-electrodes amperometric biosensor approach.* Dordrecht: Springer.
122. Midé Engineering Smart Technologies. Mide Tehnology Corp., Medford, MA. http://www.mide.com/
123. IXYS Efficiency through Technology. http://www.ixys.com
124. Colomer-Farrarons, J., P. Miribel-Català, I. Rodríguez, and J. Samitier. 2009. CMOS front-end architecture for in-vivo biomedical implantable devices (paper presented in the 35th International Conference of Industrial Electronics, IECON, Porto, Portugal, November 3–5).

3 Energy Harvesting
Thermoelectric and Microsystems Perspectives and Opportunities

Terry J. Hendricks

CONTENTS

3.1   INTRODUCTION

Most industrial and transportation processes worldwide waste 50%–70% of the fuel energy input, leading to vast amounts of wasted thermal energy that is both available and recoverable. Recent studies indicate that approximately 12.5 quads of thermal energy are available across a spectrum of transportation platforms, including light-duty vehicles (e.g., passenger vehicles, minivans, and sport utility vehicles), and heavy vehicles (e.g., class 4–class 8 trucks) in the United States alone. Recent additional studies indicate that there are another approximately 10 quads of thermal energy available across a variety of industrial processes, including aluminum, glass, steel, cement, paper and pulp, and other processes in the United States. This energy typically is dissipated to the environment in exhaust and coolant systems of these vehicles and industrial processing plants. Thermoelectric power generation (TEG) is one important technology that is available to recover this energy and convert it to useful electrical energy. TEG systems are typically quiet, low maintenance, capable of high reliability and stealthy operation when designed properly, and can transition gracefully to different power levels when necessary. TEG power technology has been used to recover waste energies in certain niche energy recovery applications (e.g., truck exhausts, wood-burning stoves) and in small combustion-driven systems (e.g., natural gas line sensors). Recent advancements in thermoelectric (TE) materials have created the potential to harness this energy and convert it at much better energy conversion efficiencies (near or greater than 10%) than in past applications. Thermoelectric energy recovery systems using new TE materials have thermal, structural, and thermoelectric design challenges to overcome in designing high-performance, robust, and flexible systems that take advantage of these new materials. Thermoelectric, thermal, and structural considerations are necessarily interdependent in these system designs and must be dealt with simultaneously to satisfy the TE system's operational requirements and achieve the high performance and robustness desired. This chapter discusses these design challenges and various design and analysis techniques used to overcome them in current and future systems.

Nomenclature

English

A	Area (m^2 or cm^2)
A_f	Total fin area (m^2 or cm^2)
C_p	Specific heat (J/kg-K)
D_h	Hydraulic diameter (m or cm)
E	Electric field intensity (V/m)
h, h_c	Convective heat transfer coefficient (W/m^2-K or W/cm^2-K)
K	Thermal conductance (W/K)
L	TE element length (m or cm)
m	External to internal electrical resistance ratio
$\dot{m}_h$	Exhaust mass flow rate (kg/s)
m_{vol}	Control volume mass (kg)
N	Number of TE couples
I	Current (amperes)
J	Current density (amperes/m^2 or amperes/cm^2)
P	Power (watt)
Pr	Prandtl number
q	Heat flow [watt]
Q	Heat flow [watt]
R	Electrical resistance (ohms) or thermal contact resistance (K/W)
Re	Reynolds number
t	Time [seconds]

T	Temperature (K)
U	Overall heat transfer coefficient (W/m²-K or W/cm²-K)
UA	Heat exchanger conductance (W/K)
V	Voltage (volt or microvolt)
x	Length position along TE element (m or cm or mm)
Z	TE material figure of merit (= $\alpha^2/\rho\kappa$) (1/K)

Greek

α	Seebeck coefficient (V/K or μV/K)
β	Heat loss factor
δ	Electrical contact resistance ratio
Δ, d	Change in
ε	Heat exchanger effectiveness
Γ	TE element geometry factor, A_{TE}/L (m or cm)
Γ_{fl}	Fluid density (kg/m³)
Γ	Heat exchanger heat transfer to required pumping power ratio
κ	Thermal conductivity (W/m-K or W/cm-K)
η	Conversion efficiency
η_F	Fin heat transfer efficiency
ρ	Electrical resistivity (ohm-m or ohm-cm)
σ	Electrical conductivity (S/m or S/cm)
μ_{fl}	Fluid viscosity (N-s/m²)

Subscripts

h	Hot side of TE device
c	Cold side of TE device
cen	Center of
ch	Channel
conn	Connector
contact	Electrical contact resistance value
cross	Cross-sectional
ex	Heat exchanger value
exh	Exhaust flow quantity
f	Fluid
int	Internal TE device resistance
loss	Heat loss quantity
max	Maximum power or efficiency
nat	Natural convection
TE	TE element or thermoelectric
o	External load resistance
s	Surface
opt	Optimum design conditions
∞, amb	Ambient
1, 2, 3	Location of the TEG/control volume in direction of flow

3.2   THERMOELECTRIC DESIGN—OPTIMIZATION AND CONSTRAINTS

Thermoelectric (TE) technology can recover and convert a portion of the waste heat from a variety of industrial, commercial, transportation, and military systems through temperature differentials between the various exhaust streams or hot environments and cooling environments available in these applications. Table 3.1 shows some typical waste exhaust stream temperatures available in various industrial and transportation applications. The US Department of Energy, Office of Energy Efficiency and Renewable Energy, has performed extensive analysis of petroleum-derived fuel use across the United States [1]. This analysis typically uses separate fuel-use categories for light-duty vehicles, minivans and sport-utility vehicles (SUVs), and medium-/heavy-duty vehicles. The transportation energy data [1] show that in 2002, for example, the fuel usage by light-duty vehicles, minivans, and SUVs represented approximately 16.27 quads of energy (1 quad = 10^{15} Btu). Approximately 35% of this energy was dissipated in the high-temperature exhaust streams of these vehicles; therefore, approximately 5.7 quads of waste thermal energy were available to recover in exhaust streams in this vehicle category in 2002. This

TABLE 3.1
Available Waste Energy Temperatures in Different Economic Sectors

Economic Sector and Applications	Available Stream Temperatures
Industrial Sector	
Water/steam boilers	~150°C
Aluminum smelting	~960°C
Aluminum melting	~750°C
Glass furnaces	485°C–1400°C
Transportation Sector	
Light-duty vehicle exhaust	350°C–600°C (depending on location)
Heavy-duty vehicle exhaust	300°C–500°C (depending on location)

Sources: Industrial sector: US Department of Energy, Industrial Technologies Program, Energetics, Inc. & E3M, Inc. 2004. Energy use, loss and opportunities analysis: US Manufacturing and Mining. Transportation sector: Hendricks, T. J., and W. T. Choate. 2006. US Department of Energy, Industrial Technology Program, http://www.eere.energy.gov/industry/imf/analysis.html; Hendricks, T. J., and J. A. Lustbader. 2002. *Proceedings of the 21st International Conference on Thermoelectrics,* Long Beach, CA, IEEE catalogue no. 02TH8657, 381–386; Hendricks, T. J. and J. A. Lustbader. 2002. *Proceedings of the 21st International Conference on Thermoelectrics,* Long Beach, CA, IEEE catalogue no. 02TH8657, 387–394; Crane, D. T., J. W. LaGrandeur, and L. E. Bell. 2010. *Journal of Electronic Materials* 39 (9): 2142–2148.

same transportation energy data [1] show that in 2002 the fuel usage by medium-/ heavy-duty vehicles was approximately 5.03 quads of energy. Approximately 30% of this energy was dissipated in the high-temperature exhaust streams of these vehicles, so another approximately 1.5 quads of waste thermal energy were available to recover in exhaust streams in this vehicle category in 2002. In 2008, the energy usage in the light-duty vehicle, minivan, and SUV sectors rose to approximately 16.4 quads and the medium-/heavy-duty vehicle sector stayed roughly the same at approximately 5.02 quads [1]. Therefore, the high-temperature waste thermal energy available in these two vehicle sectors has stayed roughly equivalent during that time period.

The US Department of Energy Industrial Technologies Program office also has estimated waste thermal energy dissipated in various industrial processes throughout the United States [2,3]. This analysis categorized and characterized waste thermal energy in industrial processes such as aluminum smelting, glass processing, steel processing, paper and pulp processing, cement processing, and several others. The analysis found considerable waste of thermal energy in these industrial processes and, therefore, highly significant opportunities for industrial waste heat recovery. Table 3.2 shows typical process efficiencies for some of these processes [2].

Tables 3.1 and 3.2 show that there are enormous amounts of industrial waste thermal energy available for recovery and that waste occurs at temperatures that are quite compatible with the temperature-dependent performance of many TE materials discussed in other chapters in this handbook. Estimates indicate there are approximately 10 quads of waste thermal energy available in various industrial processes and about 1.8 quads of this waste thermal energy are recoverable [2], meaning it is at temperatures where it could be recovered.

Waste heat in these exhaust streams and processes can be recovered either within a heat exchanger integrated with a TE device hot side tied directly into the exhaust

TABLE 3.2
Typical Process Efficiencies in Key US Industrial Processes and Applications

Industrial Process	Typical Thermal Efficiency
Paper drying	~48%
Glass processing	45%–79%
Aluminum processing	40%–65%
Cement calcination	30%–70%
Distillation column	25%–40%
EAF steelmaking	~56%
Power production	25%–44%
Steam boilers	~80%

Source: US Department of Energy, Industrial Technologies Program, Energetics, Inc. & E3M, Inc. 2004. Energy use, loss and opportunities analysis: US Manufacturing and Mining

stream (i.e., real-time recovery) or by storing it in a thermal energy storage medium for use in the future. New opportunities for efficiently recovering waste heat have been created because of advances in micro- and nanotechnologies. The key to identifying TE waste heat recovery applications is to determine when thermal exchange between existing process fluids in a given system is not an available option or provides no useful technical or economic benefit, or when electrical power generated by TE systems has a beneficial intrinsic value within the system.

Extensive system-level research has been performed in the past 5 years to design TE systems to recover and convert waste thermal energy to useful electrical energy in various industrial and transportation systems. This work has been useful in demonstrating what TE conversion efficiencies and power levels are possible with newer, advanced TE materials (i.e., skutterudites, lead–antimony–silver–telluride nanocomposites, half-Heuslers) and more conventional TE materials (i.e., bismuth telluride) in these different waste energy applications.

Designing TE generator systems for these waste heat recovery (WHR) applications typically involves two processes: *design optimization,* which follows design techniques discussed in Angrist [7] and Rowe [8], and *design performance prediction,* which follows techniques discussed in Hogan and Shih [9], Hendricks et al. [10], and Crane [11].

3.2.1 TE DESIGN OPTIMIZATION

Design optimization generally involves maximizing the TE conversion efficiency, TE power output, power density, specific power (i.e., power per mass), or other relevant design parameter for the TE system or application of interest. The TE conversion efficiency, η, is typically given by

$$\eta = P/Q_h \tag{3.1}$$

where P = power output and Q_h is the thermal energy input to the TE hot side [7,8]. The power output is generally given by

$$P = \left[\frac{(N \cdot \alpha \cdot \Delta T)^2}{(N \cdot R_{int} + R_o)^2} \cdot R_o \right] \tag{3.2}$$

where
  R_{int} and R_o are familiar internal and external load resistances, respectively
  $\Delta T = T_h - T_c$
  T_h = TE device hot-side temperature
  T_c = TE device cold-side temperature
  $\alpha = |\alpha_p| + |\alpha_n|$ for a single couple

This equation indicates the multiplier, N, the number of couples in a device, for a multiple couple system.

Optimization of conversion efficiency for a given set of TE materials and TE element geometries is discussed at length in Angrist [7] and Rowe [8] and leads to the well-known relationship:

$$\eta_{max} = \left(\frac{P}{Q_h}\right)_{max} = \left[\frac{T_h - T_c}{T_h}\right] \cdot \left[\frac{\left(1 + Z^*\overline{T}\right)^{1/2} - 1}{\left(1 + Z^*\overline{T}\right)^{1/2} + \frac{T_c}{T_h}}\right] \tag{3.3}$$

where $\overline{T} = (T_h + T_c)/2$. Recent design optimization techniques have leveraged these basic equations to develop more sophisticated analysis and optimization algorithms because, in most waste heat recovery and TE power system designs, the external exhaust and ambient temperatures, T_{exh} and T_{amb}, are usually known more clearly as requirements or bounding conditions. Consequently, it is difficult to surmise a priori what the best combination of T_h and T_c for a given TE material set will be in any given waste heat recovery application, or whether maximum efficiency conditions, maximum power, or maximum power density conditions are going to be the most appropriate system parameters to optimize. In addition, and more importantly, it is necessary to couple the TE device optimization closely with the hot-side and cold-side thermal design optimization to create high-performance TE power systems. This necessarily requires one to analyze and optimize the TE design based on the external exhaust and ambient temperatures, T_{exh} and T_{amb}, in any given application.

Hendricks and Lustbader [4,5] first began investigating this and developing more powerful optimization techniques that laid the foundation for the advanced optimization techniques discussed later. Their approach created a "system of optimization design equations" that coupled the hot-side heat transfer, Q_h, and cold-side heat transfer, Q_c, design optimization with the TE device design optimization. The analysis considers the TE system schematically represented in Figure 3.1 where a simple, multiple-couple TE system is depicted. This type of system configuration is common whether one is considering bulk elements or thin-film elements.

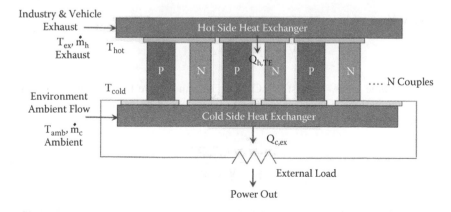

FIGURE 3.1 Industrial or vehicle thermoelectric energy recovery system schematic.

Angrist [7] and Rowe [8] discuss that the amount of heat required on the TE device hot side, Q_h, is given by

$$Q_h =$$

$$N \cdot \left[\alpha \cdot I \cdot T_h - \frac{1}{2} \cdot I^2 \cdot \left(\frac{\rho_p}{\gamma_p} + \frac{\rho_n}{\gamma_n} + 2 \cdot R_{contact,h} \right) + \left(\kappa_p \cdot \gamma_p + \kappa_n \cdot \gamma_n \right) \cdot \left(T_h - T_c \right) \right] \quad (3.4a)$$

From Equations (3.2) and (3.4a) it can be deduced with straightforward mathematics that the amount of cold-side heat dissipated, Q_c, is given by

$$Q_c =$$

$$N \cdot \left[\alpha \cdot I \cdot T_c + \frac{1}{2} \cdot I^2 \cdot \left(\frac{\rho_p}{\gamma_p} + \frac{\rho_n}{\gamma_n} + 2 \cdot R_{contact,c} \right) + \left(\kappa_p \cdot \gamma_p + \kappa_n \cdot \gamma_n \right) \cdot \left(T_h - T_c \right) \right] \quad (3.4b)$$

The current, I, and resistance, R_{int}, are given by

$$I = \frac{N \cdot \alpha \cdot (T_h - T_c)}{N \cdot R_{int} + R_o} \quad (3.5)$$

$$R_{int} = \frac{\rho_p}{\gamma_p} + \frac{\rho_n}{\gamma_n} + R_{contact} \quad (3.6)$$

and

$$\alpha = |\alpha_p| + |\alpha_n| \text{ and } R_{contact} = R_{contact,h} + R_{contact,c}.$$

The voltage of the system is given by

$$V_o = I \cdot R_o = \alpha \cdot \Delta T - I \cdot R_{int} \quad (3.7)$$

on a per-couple basis. Angrist [7], Cobble [12], and Rowe [8] show that this entire system of equations (Equations (3.2) and (3.4–3.7)) can be optimized to achieve the maximum conversion efficiency, η_{max}, when

$$\frac{\gamma_n}{\gamma_p} = \sqrt{\frac{\rho_n \cdot \kappa_p}{\rho_p \cdot \kappa_n}} \quad (3.8)$$

This is basically a TE element design requirement, and Z^* in Equation (3.2) is given by

$$Z^* = \frac{\alpha^2}{\left[\sqrt{\rho_n \cdot \kappa_n} + \sqrt{\rho_p \cdot \kappa_p}\right]^2} \tag{3.9}$$

Furthermore, the maximum efficiency with respect to external resistance is achieved when

$$m_{opt} = \left(\frac{R_o}{N \cdot R_{int}}\right)_{opt} = \sqrt{(1+\delta)^2 + (1+\delta) \cdot Z^* \cdot \overline{T}} \tag{3.10}$$

where $\delta = (R_{contact}/R_{int})$.

Hendricks and Lustbader [4,5] formulated this set of optimization equations into a set of self-consistent design optimization equations given by

$$\left[\frac{Q_h}{N \cdot \gamma_n}\right]_{opt} = f_q(T_h, T_c) \tag{3.11a}$$

$$\left(\frac{I}{\gamma_n}\right)_{opt} = f_i(T_h, T_c) \tag{3.11b}$$

$$\left(\frac{V}{N}\right)_{opt} = f_v(T_h, T_c) \tag{3.11c}$$

$$\left(\frac{\gamma_n}{\gamma_p}\right)_{opt} = \sqrt{\frac{\rho_n \cdot \kappa_p}{\rho_p \cdot \kappa_n}} = f_g(T_h, T_c) \tag{3.11d}$$

$$\eta_{max} = \left(\frac{P}{Q_h}\right)_{max} = \left[\frac{T_h - T_c}{T_h}\right] \cdot \left[\frac{m_{opt} - 1 - \delta}{m_{opt} + (1+\delta)\dfrac{T_c}{T_h}}\right] = f_\eta(T_h, T_c) \tag{3.11e}$$

These are functions of only temperatures T_h and T_c because the TE material properties generally are functions of temperature in any design optimization or design performance analysis. Recognizing the importance of hot-side and cold-side heat transfer and heat exchanger performance in heat recovery systems, Hendricks and Lustbader completed the system optimization analysis for TE systems depicted in Figure 3.1 by simultaneously including the design analysis for the hot-side and cold-side heat exchangers. Their analysis includes hot-side and cold-side heat losses by parametrically accounting for them with heat loss factors, β_{hx}, $\beta_{h,TE}$, β_{cx}, and $\beta_{c,TE}$, which are basically heat loss fractions of incoming energy to a given heat exchanger or interface. The thermal equation for the heat transfer supplied by the hot-side heat exchanger to the TE device then becomes

$$Q_{h,TE} = \frac{(T_{exh} - T_h) \cdot (1 - \beta_{h,TE})}{\left[\dfrac{1}{\dot{m}_h \cdot C_{p,h} \cdot \varepsilon_h \cdot (1 - \beta_{hx})} + R_{th,h}\right]} \qquad (3.12)$$

The thermal equation for the heat transfer dissipated by the cold-side heat exchanger from the TE device then becomes

$$Q_{c,TE} = \frac{(T_c - T_{amb})}{\left[\dfrac{(1 - \beta_{cx})}{\dot{m}_c \cdot C_{p,c} \cdot \varepsilon_c} + R_{th,c}\right] \cdot (1 - \beta_{c,TE})} \qquad (3.13)$$

The heat loss fractions are generally defined as the amount of heat loss compared to the heat loss entering a given component, whether it be within the heat exchangers themselves or the interfaces at which heat is entering or leaving the TE device hot side or cold side. Therefore, the heat loss fractions are generally defined as

$$\beta_{h,TE} = \frac{Q_{loss,h,TE}}{Q_{h,TE}} \quad \beta_{hx} = \frac{Q_{loss,h,ex}}{Q_{h,ex}} \quad \beta_{c,TE} = \frac{Q_{loss,c,TE}}{Q_{c,TE}} \quad \beta_{cx} = \frac{Q_{loss,c,ex}}{Q_{c,ex}}$$

$R_{th,h}$ and $R_{th,c}$ in Equations (3.12) and (3.13) are the sum of all thermal resistances at the interface between the TE device and the hot-side and cold-side heat exchangers, respectively. Energy balance requirements then couple these thermal transfer relationships to the TE design optimization equations in Equations (3.11a)–(3.11e) to complete the system design optimization. Figure 3.1 shows that this was basically a four-lumped-node design optimization model that simultaneously accounted for heat exchanger and TE device performance in the design optimization. This set of optimization equations specified the family of optimum TE designs, including the number of couples, p-type and n-type element areas, voltage and current output, power and maximum conversion efficiency for any given T_h and T_c, and a specific exhaust temperature, T_{exh}, and ambient temperature, T_{amb}.

This design optimization analysis is required in waste energy recovery applications because what is usually known are the exhaust flow conditions and ambient temperatures in a given waste energy recovery environment or situation. The goal is to identify the best T_h and T_c conditions at which to operate to create the maximum conversion efficiency or power output, important information about the TE design at those conditions, and performance sensitivities and trade-offs around these optimum design points. The optimum selection of T_h and T_c involves critical trade-offs in conversion efficiency, power, and hot-side and cold-side thermal transport in any given waste energy recovery application. These trade-offs are demonstrated in Figure 3.2

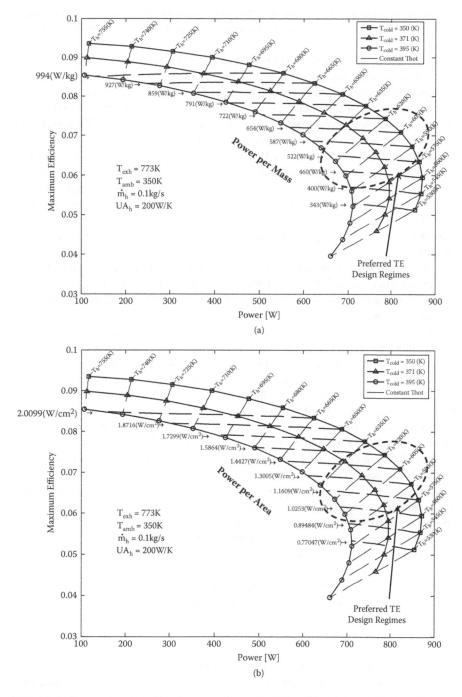

FIGURE 3.2 Maximum efficiency–power output map for typical TAGS: PbTe [21,22] TE element designs: $UA_h = 200$ W/K, $T_{exh} = 773$ K, $T_{amb} = 350$ K.

and the associated discussion. In fact, this is the case in most TE power generation analyses and applications whether they are necessarily waste energy recovery related or not.

Figure 3.2 shows typical results produced from this optimization analysis in the form of maps depicting the relationship between maximum efficiency and power (maximum efficiency–power) that can be created for any TE energy recovery system design and any set of temperature-dependent TE material properties. Two different maximum efficiency–power maps are shown: (a) one with constant specific power lines superimposed, and (b) one with constant power-per-area lines superimposed on the map. These analyses were performed using typical p-type TAGS-85 and n-type PbTe thermoelectric properties [21,22], but they can be created for any p-type/n-type TE material combination, including segmented element designs. These maps are quite powerful in the amount of information that is conveyed in one design optimization map. They are generally created for a given exhaust temperature, ambient temperature, hot-side heat exchanger UA_h (defined in Section 3.3), and exhaust mass flow rate, $\dot{m}_h$.

The main efficiency–power curves represent the loci of optimum designs having maximum efficiency and the plotted power at varying TE hot-side temperatures, T_h, and cold-side temperatures, T_c. The shape of these curves is produced directly from the coupling and interaction between the hot-side heat exchanger performance (and heat transfer) and the TE device performance in the optimization analysis described by Equations (3.11)–(3.13). As T_h increases for a given T_c, the TE device efficiency increases, but for a constant T_{exh} and UA_h the hot-side heat transfer is decreasing. Consequently, the power increases with the TE efficiency up to a point where maximum power is achieved, after which the power decreases because hot-side heat transfer decreases too much to offset the TE efficiency increase. Each optimum design along these curves has a different number of couples, N; optimum p-type and n-type element area, A_p and A_n; and current, I, as T_h and T_c combinations vary as shown in Figure 3.2. These curves first clearly demonstrate the trade-off between maximum efficiency and power for the conditions of a given exhaust stream, T_{exh}, UA_h, and $\dot{m}_h$ in typical waste energy recovery applications. The constant hot-side temperature lines help to identify where TE designs with equivalent T_h lie and their relationship to one another in efficiency and power. The efficiency–power curves tend upward and to the right (efficiency and power increasing) as either or both T_{exh} and UA_h increase in magnitude. The analytical power of these curves is that they effectively and rapidly provide the opportunity to identify and investigate multiple TE system designs throughout the potential design domain for a given waste heat recovery application.

Figure 3.2(a) also shows lines of constant specific power (power per TE device mass). The mass accounted for here is only estimates of the TE device mass, including insulating ceramics, copper interconnections straps, and the TE elements themselves. It is clear in Figure 3.2(a) that specific power generally increases as one transitions into regions of higher maximum efficiency. Figure 3.2(b) also shows lines of constant power flux (power per area), which also can be related to the hot-side heat flux because of the maximum efficiency information included in the map. It is clear that power flux increases as one moves into regions of higher maximum efficiency, which can have significant ramifications in applications where TE device miniaturization is critical.

These maximum efficiency–power maps show the potential points of optimum design and the efficiency–power sensitivities and design trade-offs available for given waste heat recovery applications. These maps can be generated for any selection of p-type and n-type TE materials discussed in other chapters of this book, and for any particular single-material element or segmented-element design. They also define "preferred TE design regimes" or preferred optimum design regions, which are superior because of (1) their higher maximum efficiency with only a small power penalty, and (2) their higher specific power performance.

Hendricks and Lustbader [4,5] also extended the design optimization techniques described earlier and, based on Equations (3.11a)–(3.11e), to segmented-element optimization using the techniques described in Swanson, Somers, and Heikes [13]. The same types of maximum efficiency–power maps discussed before can be generated in the resulting segmented-element design optimization analyses.

Additional numerical, system-level, TE design optimization has been demonstrated by Crane [14], Crane and Bell [15,16], and Crane and Jackson [17]. Constrained, nonlinear minimization functions were defined to solve the multiparameter design problems using either gradient-based, genetic algorithms or a hybrid of the two approaches. A better understanding of the interactions between various design variables and parameters can be gained from this type of multiparameter optimization approach. More than 20 different design variables, including fin and TE dimensions and dozens of different design parameters, can be optimized. A selection of different design constraints includes minimum power density, maximum hot- and cold-side pressure drops, maximum total mass, and minimum output power. Additional constraints include maximum TE surface temperatures and maximum temperature gradients across the TE elements to help improve design robustness. Choices for analysis objective function include maximum gross or net power, maximum efficiency, and maximum gross or net power density, which can be based on either total mass or TE mass. An optimization analysis can be conducted once the design variables, parameters, constraints, and objective function have been selected. The result is a nominal design that can be fine-tuned through parametric analysis on selected variables and parameters and then used in an operating model where the design conditions can vary.

3.2.2 TE Design Performance

TE system design performance is the second stage of the design process where an optimum design identified in the design optimization process is then analyzed for a variety of nominal and off-nominal design conditions. A more sophisticated multiple-node and/or multiple-volume analysis is typically employed at this stage. Analytically exact techniques have been described by Sherman, Heikes, and Ure [18] for p–n TE couple designs shown in Figure 3.1. Hogan and Shih [9] describe a typical performance analysis approach based on solving Domenicali's equation [19] for one-dimensional energy balance in the p-type and n-type elements within a device:

$$\frac{\partial}{\partial x}\left(\kappa(x)\cdot\frac{\partial T(x)}{\partial x}\right) = -\rho\cdot J^2 + J\cdot T(x)\cdot\frac{\partial\alpha(x)}{\partial x} \qquad (3.14)$$

and the equation for heat flux within the elements:

$$q(x) = J\cdot T(x)\cdot\alpha(x) - \kappa(x)\cdot\frac{\partial T(x)}{\partial x} \qquad (3.15)$$

These equations are valid in typical parallelepiped or constant diameter cylindrical TE elements used in most typical TE devices. Mahan [20] and Hogan and Shih [9] describe how Equations (3.14) and (3.15) can be reformulated into a coupled set of first-order differential equations for T(x) and q(x) within the TE element design given by

$$\kappa(x)\cdot\frac{dT(x)}{dx} = J\cdot T(x)\cdot\alpha(x) - q(x) \qquad (3.16)$$

$$\frac{dq(x)}{dx} = \rho(x)\cdot J^2\cdot\left[1 + Z(x)\cdot T(x)\right] - \frac{J\cdot\alpha(x)\cdot q(x)}{\kappa(x)} \qquad (3.17)$$

Hogan and Shih [9] originally presented Equation (3.17) as

$$\frac{dq(x)}{dx} = \rho(x)\cdot J^2\cdot\left[1 - Z(x)\cdot T(x)\right] - \frac{J\cdot\alpha(x)\cdot q(x)}{\kappa(x)}$$

This has since been shown to be a typographical error in reference 9, per discussions with Hogan (July 19, 2010) and an errata sheet that shows the correction as shown in Equation (3.17).

Hogan and Shih discuss how coupled Equations (3.16) and (3.17) can be solved for T(x) and q(x) profiles in the TE element with iterative numerical analysis techniques and appropriate boundary conditions on the cold-side and hot-side temperatures of given a TE device. The p-type and n-type elements are typically nodalized into a number of finite isothermal TE layers in the lengthwise direction (i.e., x-direction) as shown in Figure 3.3. Equations (3.16) and (3.17) are applied in a finite-difference formulation at each ith layer to yield

$$T_{i+1} = T_i + \frac{\Delta x}{\kappa_i}\cdot\left(I\cdot T_i\cdot\alpha_i - q_i\right) \qquad (3.18)$$

$$q_{i+1} = q_i + \left[\rho_i\cdot J^2\cdot\left(1 + \frac{\alpha_i^2\cdot T_i}{\rho_i\cdot\kappa_i}\right) - \frac{J\cdot\alpha_i\cdot q_i}{\kappa_i}\right]\cdot\Delta x \qquad (3.19)$$

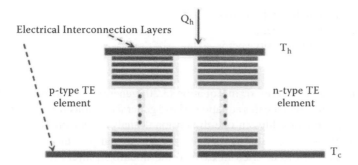

TE elements expanded to show and emphasize layered nodalization

FIGURE 3.3 Typical isothermal layer nodalization used in TE elements in iterative numerical techniques in solving Equations (3.16) and (3.17).

This formulation corrects the typographical error in reference 9 per discussions with Hogan (July 19, 2010) and an errata sheet created. This technique does not necessarily couple the q(x) profile with hot-side and cold-side heat exchanger performance, but it does provide useful T(x) and q(x) information for any given T_h and T_c conditions of interest and provides required hot-side heat transfer, Q_h, and cold-side heat transfer, Q_c, that the heat exchangers must accommodate.

In order to truly perform a system-level design performance analysis, the solution to these differential equations must be coupled with hot-side and cold-side heat exchanger analysis. Hendricks et al. [10] have recently used thermoelectric analysis routines within the ANSYS® version 12.0 analysis software package, which solve relationships similar to Domenicali's equation (Equation (3.14)) within a TE device design subject to appropriate thermal boundary conditions, to perform this type of system-level design performance analysis. This type of analysis relies on characterizing the UA_h of the hot-side heat exchanger design using a given hot-side exhaust temperature and mass flow rate entering the heat exchanger design and details of the heat exchanger structure and interfaces. The ANSYS® version 12.0 software analysis code generally allows one also to nodalize the TE elements as shown in Figure 3.3; it uses variational principles and finite volume techniques applied to the conservation of energy and continuity of electric charge to solve a set of simultaneous equations for heat flow, current, and temperatures within the TE elements under transient or steady-state conditions [23]. Under steady-state conditions, these relationships in their most general, three-dimensional form are given by

$$\nabla \cdot \left(T \cdot [\alpha] \cdot \vec{J} \right) - \nabla \cdot \left([\kappa] \cdot \vec{\nabla} T \right) = \dot{q} \tag{3.20}$$

$$\nabla \cdot \left([\sigma] \cdot [\alpha] \cdot \vec{\nabla} T \right) - \nabla \cdot \left([\sigma] \cdot \vec{E} \right) = 0 \tag{3.21}$$

where
   $[\alpha]$ = Seebeck coefficient matrix
   $[\sigma]$ = electrical conductivity matrix

[κ] = thermal conductivity matrix
J̄ = current density vector
Ē = electric field intensity vector
T = temperature

The heat generation term, q̇, in Equation (3.20) generally contains the $\rho \cdot J^2$ heat dissipation term in Domenicali's equation. These general relationships converge down to Domenicali's relationships in one-dimensional analyses depicted in Figure 3.3. The power of the ANSYS® analysis is that it allows any temperature-dependent thermoelectric properties to be used in each of the nodalized layers in Figure 3.3, so that it is straightforward to model segmented element designs in a system design. Furthermore, ANSYS® 12.0 theoretically allows one to couple the solution of Equations (3.20) and (3.21) to hot-side and cold-side heat exchanger performance models to produce system-level predictions.

The results of such an analysis produce TE module efficiency–power maps similar to those shown in Figure 3.4. These maps show the module efficiency–power for various external load resistances and temperature differentials that are created across the TE module directly because of the interaction and coupling of the TE device and hot-side and cold-side heat exchanger performance. The external resistance ratios $(R_o/N \cdot R_{int})$ and temperature differentials $\Delta T = (T_h - T_c)$ are superimposed on the

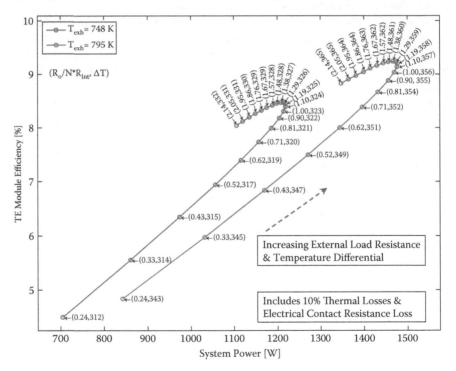

FIGURE 3.4 Typical TE module efficiency-power maps for varying external resistance and current.

module efficiency–power curves in Figure 3.4. The analysis includes the effect of interface contact resistance and parasitic thermal losses. It is clear in the Figure 3.4 results how temperature differential, $\Delta T = (T_h - T_c)$, increases across the TE module as the external load resistance increases. This is the direct result of the trade-off between system voltage and current and TE heat flows, power, and conversion efficiency as the external resistance is varied.

The analysis in Figure 3.4 was performed for exhaust flow temperature, T_{exh}, from 780 to 733 K, exhaust mass flow rate, $\dot{m}_h$, of 0.16 kg/s, and a UA_h of approximately 377 W/K; however, the analysis can be performed at various exhaust flow conditions, T_{exh}, exhaust mass flow rate, $\dot{m}_h$, and heat exchanger performance characterized by the UA_h. Figure 3.4 also shows, for example, a typical impact of reducing the exhaust temperature, T_{exh}, from 795 to 748 K alone on the power output and module efficiency. The benefit of depicting the analysis results in this manner is that the maximum efficiency and maximum power points, and their relationship to one another and other nonoptimum performance points, in the overall potential operating performance space can be quickly and clearly seen. Critical design performance decisions concerning various operating points then can be made quickly and intelligently with regard to what the entire design performance envelope looks like.

One noteworthy and clearly identified performance characteristic is that the point of maximum power occurs at an external resistance condition of $R_o > N \cdot R_{int}$ in this type of design performance analysis, because it is performed at constant T_{exh} and T_{amb} conditions and not constant T_h and T_c conditions. In fact, Figure 3.4 demonstrates clearly that ΔT does not stay constant along the module efficiency–power curves. Hendricks et al. [10] discuss the reasons for this, which are directly tied to the TE power output relationship in Equation (3.2). Determining the external resistance that defines maximum power point when ΔT is not constant leads to the following relationship described in Hendricks et al. [10]:

$$R_{o,opt} = N \cdot R_{int} + \frac{2 \cdot R_{o,opt} \cdot \left(N \cdot R_{int} + R_{o,opt}\right)}{\Delta T} \cdot \frac{\partial(\Delta T)}{\partial R_o} \qquad (3.22)$$

This relationship clearly shows that the optimum external resistance that maximizes power is greater than $N \cdot R_{int}$. In fact, this equation can ultimately be cast into a quadratic relationship for $R_{o,opt}$, which can be explicitly solved for $R_{o,opt}$ in any given design performance analysis.

Figure 3.4 also clearly shows that the maximum TE module efficiency condition occurs at $R_{o,opt}$ value where

$$\left(\frac{R_{o,opt}}{N \cdot R_{int}}\right)_{efficiency} > \left(\frac{R_{o,opt}}{N \cdot R_{int}}\right)_{power} > 1 \qquad (3.23)$$

which are both greater than 1. However, the maximum module efficiency–power point can be quite close in magnitude to the maximum power condition. Therefore, in waste heat recovery applications and other TE power generation applications where

ΔT is not constant, one must not assume that the $[R_{o,opt}/(N \cdot R_{int})] = 1$ condition applies. Each particular design performance analysis must establish these operating points and their relationship to one another in any given waste heat recovery application.

3.2.3 Sectioned TE Design Optimization

Table 3.1 shows that there are many waste heat recovery applications where available exhaust gases and exhaust streams could be at high temperatures and, therefore, provide large temperature differentials across TE waste heat recovery devices. This is especially true if effective thermal cooling techniques can keep cold-side temperatures low. Large available temperature differentials create the opportunity to utilize a relatively advanced TE design technique called "sectioned design." With high exhaust stream temperatures and therefore large temperature differentials, the exhaust flow through the hot-side heat exchangers in a TE waste heat recovery system can undergo relatively large temperature drops along the flow length as it transfers its thermal energy to the TE device hot side. In addition, the hot-side thermal transfer decreases significantly along the flow length. Consequently, the TE devices within the design can experience large differences in hot-side temperature and hot-side thermal transfer as the exhaust flow temperature decreases. It is impossible to design one optimal TE device that operates at maximum performance levels as hot-side temperature and thermal flows significantly decrease in these situations. One would therefore like to create optimum TE device designs that accommodate this change in hot-side temperature and hot-side thermal transfer as they decrease with flow length. Sectioned design allows this type of TE device optimizing as the exhaust flow conditions change with flow length. Figure 3.5 schematically shows a typical example of this type of "dual-sectioned TE design" approach where the exhaust flow temperatures decrease significantly along the flow length, and the TE hot-side temperatures in each of the two sections are quite different.

Any number of sections can be added to satisfy given design requirements in any waste heat recovery application, but the actual measurable performance benefits gained are dependent on the magnitude of the overall $(T_{exh} - T_{amb})$ temperature differential available. Cost considerations and system complexity issues often constrain the maximum number of sections to about three. This was the number of sections described by Crane, LaGrandeur, and Bell [6]. Each TE section was optimized for a particular temperature and heat flux range in the direction of gas flow. This helps

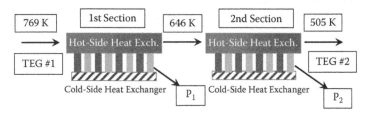

FIGURE 3.5 Typical dual-sectioned TE system design.

avoid TE incompatibility in the direction of fluid flow [15]. The TE section nearest the hot gas inlet comprised TE couples made of two-stage segmented elements to account for the large temperature gradient and heat flux between the hot and cold heat exchangers. Located axially downstream of the high-temperature TE section in the direction of gas flow, the medium-temperature TE section also is made up of two-stage segmented elements of the same materials as the high-temperature TE section. However, these elements are thinner than the elements in the high-temperature TE section due to the lower heat flux and temperature gradient. Segmented elements are not required for the lowest temperature TE section. Each of these TE sections can be operated on one electrical current or on different electrical circuits, allowing for more optimal current densities per TE section.

The design optimization and design performance analysis techniques discussed can generally be used to design each of the sections. However, there are unique and complex design trade-offs associated with the amount of heat transferred, the conversion efficiency, and power generated in each section. These must be fully evaluated in any TE energy recovery system design to achieve the optimum performance. Figure 3.6 illustrates some of these trade-offs in a simple dual-sectioned design shown in Figure 3.5. The dual-sectioned design creates significant design trade-offs between power output and efficiency in each section when seeking the optimum overall system performance, whether that is maximum overall system efficiency or maximum system power output. There are two sets of maximum efficiency–power curves in Figure 3.6; one for section 1 designs using lead–antimony–silver–telluride (LAST) materials and a second for section 2 designs using bismuth telluride materials in Figure 3.5. The maximum efficiency–power curves identify the loci of maximum efficiency designs defined by techniques in Equations (3.11)–(3.13). The resulting maximum efficiency–power maps in each section are produced by the coupled interdependence of the TE device design and hot-side and cold-side heat exchangers in sections 1 and 2.

Figure 3.6 illustrates a tremendous amount of design optimization information for the two sections of this dual-sectioned design. The section 1 efficiency–power curves show the various maximum efficiency–power points for several different hot-side and cold-side temperature combinations $(T_{h,1}, T_{c,1})$ resulting from the coupled interaction between the TE device design and hot-side heat exchanger design characterized by a $UA_h = 375$ W/K. These curves show the range and domain of efficiency and power combinations possible in section 1, with inevitably different TE couples and p-type and n-type TE areas at each design point on the curve. The design analysis does identify the TE couples and TE areas required at each design point, although this information is not explicitly shown on the curves. The hot-side heat exchanger design and its performance at each $T_{h,1}$ along the section 1 efficiency–power curves then produce unique entrance exhaust temperatures to section 2 (shown in Figure 3.5). This situation creates a family of maximum efficiency–power curves for the section 2 design, one for each of the section 1 $T_{h,1}$ conditions along the section 1 curves. Figure 3.6 shows four such section 2 design curves corresponding to four selected design points on the section 1 curve for a cold-side temperature condition of 312 K.

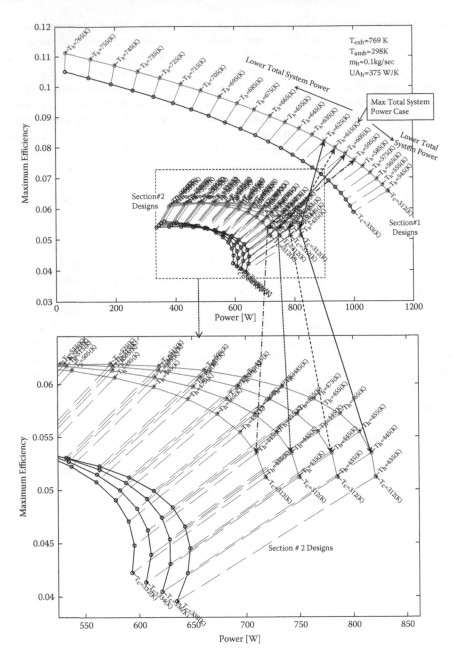

FIGURE 3.6 Typical dual-sectioned analysis results for $T_{exh} = 769$ K, $\dot{m}_h = 0.1$ kg/s, $UA_h = 375$ W/K, $T_{amb} = 298$ K.

As the section 1 design changes, efficiency and power output varies and, therefore, impacts input exhaust temperature on section 2. There is a uniquely defined section 2 maximum efficiency–power curve that dictates its efficiency–power characteristics resulting from the coupled performance of the TE device designs and the hot-side heat exchanger design in section 2. The four section 1 points selected for $T_{c,1} = 312$ K demonstrate that, as section 1 power decreases and efficiency increases, the section 2 power output generally increases while the section 2 efficiency stays roughly the same. This section 2 efficiency is largely governed by the TE material properties for bismuth telluride materials at the temperatures used in section 2. Different section 2 TE materials and properties would exhibit different efficiency–power sensitivities and performance.

This behavior and interactions between sections 1 and 2 create a maximum total system power point because the increases in section 2 power only offset or over-ride the section 1 decreases in power up to a point. The maximum total system power point generally does not occur even close to the maximum point for section 1. Understanding this design power trade-off and knowing where the maximum total system power point resides are critical to designing a dual-sectioned system to satisfy any given efficiency–power requirement, whether maximum efficiency or maximum power is of paramount interest in a given energy recovery application. This behavior and interaction between section 1 and section 2 designs can occur in general at any common cold-side temperature ($T_{c,1}$ and $T_{c,2}$) conditions, as shown in Figure 3.6, and in general for any combination of TE materials in the two sections. Figure 3.6 also shows the maximum efficiency–power reduction that occurs as cold-side tempera-tures increase (black lines for sections 1 and 2). In both of these cases, the maximum total power point may shift as a result of the impact of TE material performance or cold-side temperature effects on power output and it is crucial to quantify this shift if it occurs. These design interactions become more complex as more sections are added in a multiple-sectioned design.

3.2.4 Transient Design Performance

Steady-state models discussed earlier give an effective means to choose a nominal design point and optimize the design for a particular set of operating conditions. However, a thermoelectric generator in waste heat recovery applications may see a variety of operating conditions, which change frequently as a function of time. A primary example of this application is exhaust heat recovery in an automobile or heavy truck. In this case, a TEG is integrated into a car or truck exhaust stream and the exhaust temperature and mass flow conditions result from the transient engine load conditions dictated by the particular drive cycle that the vehicle experiences. Figure 3.7 shows an example of the temperatures and mass flows downstream of the catalytic converter in the exhaust system that are produced in a new European drive cycle (NEDC), a common automotive drive cycle used in European countries.

In order to model the TEG in different drive cycles and other dynamic operating conditions, steady-state models for TE couples and devices are defined first. Energy

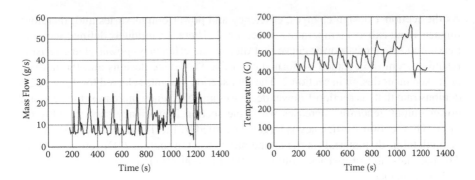

FIGURE 3.7 Time-dependent exhaust gas mass flow and temperature downstream of the catalytic converter for an inline 6 cylinder, 3.0 L displacement engine operating on the new European drive cycle (NEDC).

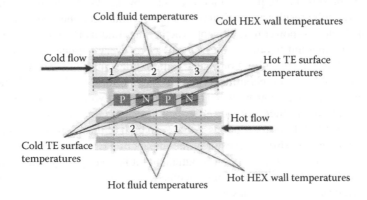

FIGURE 3.8 Schematic of TE subassembly with heat exchangers (HEXs) showing model temperature locations.

balance equations are described by Crane [11] as follows with locations shown schematically in Figure 3.8:

$$Q_{h1} + \frac{1}{2} I^2 R_{conn,h} - UA_{TE-conn}(T_{cen,h} - T_{h1}) = 0 \qquad (3.24)$$

$$Q_{h2} + \frac{1}{2} I^2 R_{conn,h} - UA_{TE-conn}(T_{cen,h} - T_{h2}) = 0 \qquad (3.25)$$

$$UA_{TE-conn}(T_{cen,h} - T_{h1}) + UA_{TE-conn}(T_{cen,h} - T_{h2}) - UA_{cross,conn}(T_{sh2} - T_{cen,h}) = 0 \quad (3.26)$$

$$hA_h(T_{fh} - T_{sh2}) - UA_{cross,conn}(T_{sh2} - T_{cen,h}) - hA_{nat}(T_{sh2} - T_\infty)$$
$$+ UA_{cross,ch,h,1-2}(\Delta T_{sh1}) - UA_{cross,ch,h,2-3}(\Delta T_{sh2}) = 0 \tag{3.27}$$

$$m_h C_{ph} \Delta T_{fh} - hA_h(T_{fh} - T_{sh2}) = 0 \tag{3.28}$$

where

$$Q_{h,i} = \alpha \cdot I \cdot T_{h,i} + K \cdot \Delta T_{TE} - \frac{1}{2} \cdot I^2 \cdot (R_{TE} + 2R_{contact}) \tag{3.29}$$

Equations (3.24) and (3.25) represent conductive heat transfer from the TE elements into their connectors. Equations (3.26) and (3.27) represent conductive heat transfer from these connectors through the fluid-carrying channel wall by way of the fins to fluid convective heat transfer connectors. They include the losses due to natural convection and radiation (Equation 3.29). is an energy balance equation for convective heat transfer into the fluid. is the standard equation for thermoelectric heat flow in power generation. The model solves these governing equations simultaneously for steady-state temperatures at each node in the direction of flow. The number of simultaneous equations varies with the number of TE elements in the direction of fluid flow.

These energy balance equations were then translated into differential equations based on Equation (3.30) and integrated into the S-function template of MATLAB®/Simulink®:

$$m_{vol} C_p \frac{dT}{dt} = Q_1 - Q_2 \tag{3.30}$$

The ($m_{vol} C_p$) term in Equation (3.30) represents the thermal mass of each control volume. This could be a TE element, TE connector, heat exchanger, or fluid thermal mass depending on the control volume. The direction of heat flow is important to make sure that the signs for Q_1 and Q_2 are correct. Otherwise, the differential equations cannot be solved correctly.

A baseline for the transient model is the optimized design from the steady-state model. Inputs for the model are similar to those of the steady-state model. Operating condition inputs include the hot- and cold-side inlet temperatures and flows and external electrical load resistance. External electrical load resistance can be set equal to the internal resistance of the TEG or it can be set at a particular constant external load. A simulated electrical load controller can be attached to the model as an additional Simulink block in order to model the effects of a varying electrical load that is not necessarily optimal. Outputs for the model are again similar to those of the steady-state model.

The model can be operated in a stand-alone mode as is or the S-function can be cut and pasted into a larger systems-level model. BMW and Ford have both used versions of this model in their larger automotive systems-level models [6]. The model can be run using single hot-side inlet flow and temperature conditions or using the hot-side inlet flow and temperature conditions for a drive cycle.

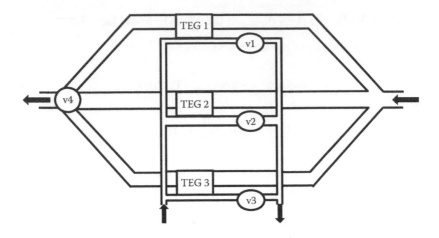

FIGURE 3.9 Schematic of multiple parallel section TEG (V stands for valve).

Additional systems-level attributes have been added to the transient model to aid in its use as part of a larger system. A maximum hot inlet temperature can be defined to prevent TE element overheating or overheating of any other part of the TEG device. A maximum hot flow can be defined to prevent excessive backpressure in the system, which can reduce engine performance if the TEG is integrated into the exhaust system of a vehicle. In addition, to better match the thermal impedance of a dynamic thermal system as defined in Crane and Bell [16], the TEG can be broken into a number of TE sections in parallel (see Figure 3.9) as opposed to sections in a series as described earlier. Having multiple TE sections can allow the TEG to operate better at low flows when the design has been optimized for higher flow rates as each of the parallel sections can be operated together or in various advantageous combinations.

Figure 3.10, a transient simulation example, shows the time-dependent power output, hot-side fluid, and TE surface temperatures for the TEG being driven toward steady-state by a set of constant operating conditions from an initial set of conditions. The temperature curves in these figures are at different axial positions on the TEG in the direction of flow as a function of time. The colder the temperatures are, the farther they are from the TEG inlet. Figure 3.11, another transient simulation example, shows efficiency and power output along with the TE surface and hot-side fluid temperatures for the TEG for an NEDC automotive drive cycle. The importance of these analysis results is that they quantify and highlight the response time of this system design in this energy recovery application. This is an important design metric that is impacted by a specific system's component weights and volumes as well as the materials and fluids used throughout the system design. System transient response is also governed by other design parameters, such as the exhaust temperature and mass flow rate and ambient environmental conditions, and it requires knowledge of the specific heats and density of each material and fluid used throughout the design. It must be characterized for each specific design and for each specific set of operating conditions, boundary conditions, and environments anticipated or of interest in a

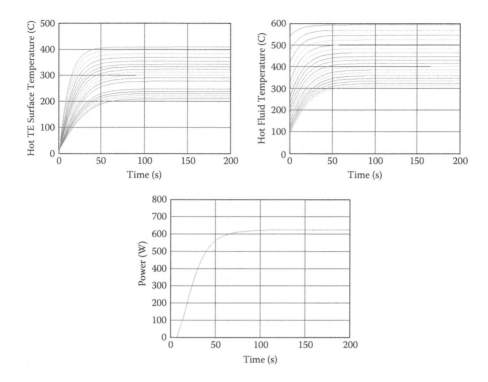

FIGURE 3.10 Simulated outputs for the transient TEG model using constant operating conditions. Temperature graphs show temperatures as a function of time at different axial positions along the TEG in the direction of fluid flow. The hottest temperature curves are nearest to the TEG hot gas inlet, while the coldest are nearest to the TEG hot gas outlet.

given energy recovery application. The amount of work required to produce credible and accurate transient response analyses should not be underestimated.

3.3 THERMAL SYSTEM DESIGN AND CONSIDERATIONS IN THERMOELECTRIC SYSTEMS

3.3.1 THERMAL SYSTEM DESIGN RELATIONSHIPS

Many waste energy recovery applications in today's environment require solutions that are compact, environmentally friendly, quiet, vibration free, without ozone-impacting fluids, and highly reliable with few or no moving parts. Advanced thermoelectric energy recovery and conversion systems envisioned in the future for waste energy recovery applications generally require that a temperature differential be maintained across the TE device while the required thermal energy is transferred in/out of the system. In any event, nearly isothermal interfaces are required on the hot and cold sides of the TE device to achieve predictable maximum performance conditions. Therefore, heat exchanger configurations that transfer the thermal energy in or out of the device must be capable of providing these nearly isothermal conditions on the interfaces with the TE conversion device and operating under their

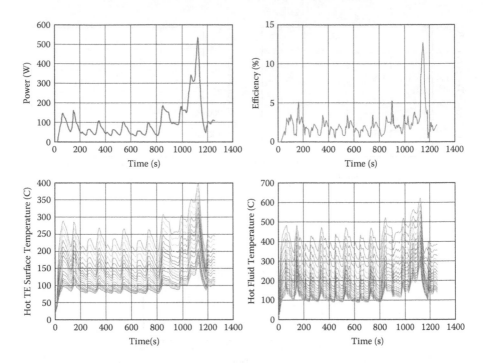

FIGURE 3.11 Simulated outputs for the NEDC automotive drive cycle using an optimized TEG from the steady-state model. Temperature graphs show temperatures as a function of time at different axial positions along the TEG in the direction of fluid flow. The hottest temperature curves are nearest to the TEG hot gas inlet, while the coldest are nearest to the TEG hot gas outlet.

influence. Furthermore, these heat exchangers will have significant weight and volume requirements on their design to assist in minimizing overall heat recovery system weight and volume.

One must realize the importance of heat transfer at both the hot side and cold side of any TE energy recovery system. A simple rewrite of Equation (3.1) (and Equation (3.3) for maximum efficiency conditions) demonstrates the relationship between TE device efficiency and these hot-side and cold-side (and parasitic loss) thermal transfers:

$$\eta = \frac{P}{Q_h} = \frac{Q_h - Q_{loss} - Q_c}{Q_h}. \tag{3.31}$$

This simple relationship focuses on the thermal transfers involved in creating the power from the TE system.

Hendricks [25–27] describes the use of the ε–NTU (number of transfer units) analysis method [20] for various basic heat exchanger configurations, counterflow, parallel flow, cross flow, and parallel counterflow. It is clear from Kays and London [24] that heat exchanger effectiveness, ε, is generally expressed by the following relationship for any flow configuration:

$$\varepsilon = \varepsilon[\text{NTU}, \left(\frac{C_{\min}}{C_{\max}}\right), \text{flow configuration}].$$ (3.32)

The ε relationship in Equation (3.32) is generally a rather complex one for the various heat exchanger flow configurations. In a heat exchanger that is providing or creating a nearly isothermal interface on one side as it transfers thermal energy, the ratio

$$\left[\frac{C_{\min}}{C_{\max}}\right] \rightarrow 0$$ (3.33)

and the Equation (3.32) relationship generally simplifies to

$$\varepsilon = 1 - \exp\left[\frac{-UA}{C_{\min}}\right].$$ (3.34)

In the advanced waste energy recovery and conversion systems considered herein, it is highly desirable that any heat exchangers coupled with the advanced TE energy conversion system should satisfy the nearly isothermal interface condition and should therefore closely follow the ε relationship in Equation (3.34). Heat exchanger design techniques and "TE design sectioning" techniques are usually employed to help ensure this condition as closely as possible. UA, the overall heat transfer conductance times the heat transfer surface area, and $C_{\min}$ are defined by Kays and London [24]. The UA_h or UA_c of typical hot-side or cold-side heat exchanger designs is evaluated by the techniques shown in Kays and London [24] and typically given by

$$UA_h = \left(\frac{1}{\sum_i R_{th,h,i}}\right) \text{ or } UA_c = \left(\frac{1}{\sum_i R_{th,c,i}}\right).$$ (3.35)

where the thermal resistances $R_{th,h,i}$ and $R_{th,c,I}$ account for all the convective thermal resistance effects of the flow and all the conductive thermal resistance effects of the heat exchanger structures. Evaluation of the thermal resistances $R_{th,h,i}$ and $R_{th,c,I}$ is therefore specific to any given hot-side and cold-side heat exchanger designs and the application. The quantity $C_{\min}$ is given by $C_{\min} = \dot{m}_h \cdot C_{p,h}$ or $C_{\min} = \dot{m}_c \cdot C_{p,c}$, depending on whether one is considering the hot-side heat exchanger or the cold-side heat exchanger, respectively.

UA_h and UA_c (Equation 3.35) are used in Equations (3.12) and (3.13) to determine the hot-side and cold-side heat transfers on the TE devices. In general, UA_h and UA_c are dependent entirely on the heat exchange fluids used, flow conditions, and the structure of the heat exchanger in any given WHR application. There can be a wide variety of heat exchanger structures and they can demonstrate some rather unique and innovative concepts. Figure 3.12 shows an example of a particularly simple and common design known as a flat-plate, parallel-extended-fin design that will be used for illustrative purposes only. The UA of this type of structure is dependent on the

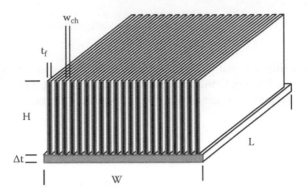

FIGURE 3.12   Common flat-plate, parallel-extended-fin heat exchanger structure.

heat transfer coefficient in the flow channels and the thermal conductance of the structure itself. The UA of this simple structure is given by

$$UA = \cfrac{1}{\left[\cfrac{1}{\eta_F \cdot h_c \cdot A_f} + \cfrac{\Delta t}{\kappa_{st} \cdot W \cdot L} \right]},$$ (3.36)

where h_c = channel heat transfer coefficient, η_F = fin thermal transfer efficiency, A_f = total fin and base heat transfer area, and κ_{st} = thermal conductivity of the plate structure (and usually the fins themselves).

The thermal conductance term in Equation (3.36) is generally straightforward to evaluate and quantify as is the fin efficiency and fin area. However, the heat transfer coefficient, h_c, is generally a function of channel Reynolds number, Re, and the Prandtl number, Pr, of the fluid [28]. There are a number of different heat transfer correlations of the following form:

$$h_c \propto \frac{\kappa_{fl}}{D_h} \cdot Re^m \cdot Pr^n$$ (3.37)

where D_h = channel hydraulic diameter, κ_{fl} = fluid thermal conductivity, and m ~ 0.5–0.8 and n ~ 0.33. Therefore, the heat transfer coefficient, h_c, is dependent on the fluid properties and the channel velocities and, therefore, the mass flow rate through the heat exchanger, as well as the channel dimensions, w_{ch} and H. The Reynolds number is generally given by

$$Re = \frac{\gamma_{fl} \cdot V_{ch} \cdot D_h}{\mu_{fl}}$$

where V_{ch} = fluid channel velocity, γ_{fl} = fluid density, and μ_{fl} = fluid viscosity.

In addition, the heat transfer coefficient is not constant throughout the flow length, generally starting out high at the flow entrance and decreasing with flow length. This is another important motivation for sectioned TE designs discussed in Section 3.2 because the actual heat transfer into the TE device hot side can decrease with flow length. Therefore, one generally desires to have slightly different TE device design along the flow length to truly optimize system performance as the UA_h decreases with flow length. Consequently, even this simple flow arrangement has some subtle and sometimes difficult challenges in evaluating the heat exchanger UA. There are many other heat exchanger configurations that are much more complex than this; the evaluation of their UA values is much more complex and often requires detailed thermal models. It is not possible to discuss how to evaluate the UA of every possible heat exchanger design option in this work. Suffice it to say that it is imperative to analyze the specific convective heat transfer conditions and environments and the conductive heat transfer of the heat exchanger structure properly when evaluating various heat exchanger design options in any given TE WHR system design.

In waste heat recovery applications, the heat exchanger fluids used vary from different gases (i.e., air, nitrogen, helium) to different liquids (i.e., water, water–glycol mixtures, oil). Many of these gases are major constituents of typical waste heat exhaust flows. The fluids' properties can vary significantly and are critical to the UA evaluation and heat exchanger performance as discussed earlier. Generally, the UA of heat exchangers using gases is low due to their generally low thermal conductivity, low Pr, and Equation (3.37) relationships, which make the first term in the denominator of Equation (3.36) dominate the UA evaluation. The UA of heat exchangers using liquids can be quite high because of their high thermal conductivity and high Pr; therefore, the first term in the denominator of Equation (3.36) becomes quite low. Table 3.3 shows the typical values for common heat exchanger fluids in WHR applications and provides a good comparison between typical gas and liquid properties. These properties and the resulting heat exchanger UA are critical factors in determining what fluids to select in WHR applications because these directly determine the temperature differential, $\Delta T = T_h - T_c$, across the TE device and therefore the TE device performance. Hendricks [26,27] presents the representative UA values of some heat exchanger designs with different fluids, but the UA evaluation is generally quite application specific using the relationships discussed previously. It is not possible to quote specific UA values for all the different and innovative gas and liquid heat exchanger options and configurations in WHR applications, but for a given heat exchanger volume, the UA of liquid heat exchangers is generally an order of magnitude or more higher than the UA of gas heat exchangers.

In detailed system modeling, whereby the heat exchanger can be broken up or nodalized lengthwise with sufficient refinement, the UA can get small enough that effectiveness in Equation (3.33) can be approximated with

$$\varepsilon = 1 - \exp\left[\frac{-UA}{C_{min}}\right] \approx \frac{UA}{C_{min}} = \frac{UA_h}{\dot{m}_h \cdot C_{p,h}} \text{ or } \frac{UA_c}{\dot{m}_c \cdot C_{p,c}} \qquad (3.38)$$

TABLE 3.3
Typical Fluid Properties of Common Gases and Liquids

Temperature (K)	κ_{fl} (W/m-K)			γ_{fl} (kg/m³)			$\mu_{fl} \times 10^5$ (N-s/m²)			Pr		
	300	500	800	300	500	800	300	500	800	300	500	800
Gases												
Air	0.0263	0.0407	0.0573	1.161	0.696	0.435	1.85	2.70	3.70	0.707	0.684	0.709
Carbon monoxide	0.0250	0.0381	0.0555	1.123	0.674	0.421	1.75	2.54	3.43	0.730	0.710	0.705
Carbon dioxide	0.0166	0.0325	0.0551	1.773	1.059	0.661	1.49	2.31	3.37	0.766	0.725	0.716
Nitrogen	0.0259	0.0389	0.0548	1.123	0.674	0.421	1.78	2.58	3.49	0.716	0.700	0.715
Helium	0.152	0.220	0.304	0.163	0.0975	—	1.99	2.83	3.82	0.68	0.668	—
Liquids												
Water	0.613	0.642	—	997.0	831.3	—	85.5	11.8	—	5.83	0.86	—
Ethylene-glycol	0.252	—	—	1114	—	—	1570	—	—	151	—	—
Engine oil	0.145	0.132	—	884.1	806.5	—	48600	470	—	6398	88	—
		(430 K)			(430 K)			(430 K)			(430 K)	

Source: Incropera, F. P., and D. P. Dewitt. 1990. *Fundamentals of heat and mass transfer,* 3rd ed. New York: John Wiley & Sons.

Equations (3.12) and (3.13) then reduce to

$$Q_{h,TE} = \frac{(T_{exh} - T_h) \cdot (1 - \beta_{h,TE})}{\left[\dfrac{1}{UA_h \cdot (1 - \beta_{hx})} + R_{th,h} \right]} \tag{3.39}$$

$$Q_{c,TE} = \frac{(T_c - T_{amb})}{\left[\dfrac{(1 - \beta_{cx})}{UA_c} + R_{th,c} \right] \cdot (1 - \beta_{c,TE})} \tag{3.40}$$

It is also quite important to evaluate the pressure drops associated with convective heat exchangers as one endeavors to increase the UA by modification of heat exchanger channel dimensions and overall geometry. The pressure drop through the heat exchanger is generally given by relationships of the form

$$\Delta P_{ex} \propto f(\mathrm{Re}) \cdot \gamma_{fl} \cdot V_{ch}^2 \tag{3.41}$$

where $f(\mathrm{Re})$ is the friction factor. There is a close relationship between the heat exchanger UA and its pressure drop. As channel dimensions decrease, the heat transfer coefficient increases through Equation (3.37) because the flow channel velocities generally increase for a given mass flow rate and the hydraulic diameter decreases. However, the pressure drop and, therefore, pumping power also increase as the channel dimensions decrease. It has been shown that there are actually optimum values of UA as a function of heat exchanger pressure drop [26,27]. This UA–pressure drop optimization is application specific and must be quantified along with optimizing TE device performance to maximize overall TE system performance in any given WHR application. It is crucial that these optimization processes be performed in an integrated and strongly coupled manner in the overall TE system design. This is why analyses in Figures 3.2 and 3.6 are so critical in the TE system design optimization.

3.3.2 MICROTECHNOLOGY HEAT EXCHANGERS IN TE SYSTEMS

In the energy recovery applications found in the transportation sector and industrial processing sector, there is generally a large amount of waste energy available to be captured and converted. This can range from 10s of kilowatts to megawatts of available thermal energy. There is also a general requirement to develop compact, lightweight, and low-cost energy conversion and heat exchange systems in any WHR design application. Therefore, in almost all applications the heat exchange systems must satisfy a general requirement to transfer very high heat fluxes (i.e., 10–100 W/cm²) across nearly isothermal interface conditions. This is particularly true of the unique energy conversion systems using advanced TE conversion materials envisioned in the next 5–10 years.

Microchannel heat exchangers are one heat transfer technology that is capable of providing this high performance. Their high performance is due to their small channel sizes, typically 10 μm to 1 mm for w_{ch} in Figure 3.12, which thereby creates very high heat transfer coefficients relative to conventional larger channels due to the Equation (3.37) relationship. Heat transfer coefficients can be approximately 400–1000 W/m²-K in air and 4000–10,000 W/m²-K in water, for example. Liao and Zhao [29] reported heat transfer coefficients of 2500–5000 W/m²-K in microchannel designs using supercritical carbon dioxide. These high heat transfer coefficients can create high UA values in gas and liquid microchannel heat exchanger designs, while the flow remains under laminar flow conditions in most microchannel flows. This laminar flow feature allows one to decouple the heat transfer augmentation from pressure drop effects in the microchannel design process. This, in turn, creates the opportunity to achieve the heat transfer augmentation while controlling increases in pressure drop through innovative designs.

Hendricks [25–27] discusses the usefulness of microtechnology or microchannel heat exchanger (MHEX) designs in satisfying future TE system requirements. There will be increasing demands to minimize weight and volume in various waste heat recovery applications as thermoelectric technology and systems evolve in the future. The TE devices themselves will need to be lighter and more compact, which will necessarily require them to accommodate and dissipate higher heat fluxes on the hot and cold sides, respectively. Higher performance heat exchangers, with higher UA values and higher heat exchange effectiveness and therefore higher heat flow capability dictated by Equations (3.12) and (3.13), will be required to allow these more compact TE devices to achieve their full performance potential. Microtechnology heat exchangers can produce the required higher UA and effectiveness levels and satisfy thermal transfer requirements with lower weight and lower volume designs (typically 1/5 to 1/10 of the weight and volume of comparable macrochannel designs) with the same thermal duty and mass flow rates.

3.3.2.1 Microtechnology Heat Exchanger Design Challenges

There are several design challenges associated with integrating MHEXs with advanced TE devices in satisfying TE system requirements. In addition to satisfying the basic TE optimization criteria, either maximum efficiency or maximum power, any optimum design must satisfy the following five additional criteria:

1. Interfacial heat flux matching requirements at both the TE device hot and cold sides
2. TE power maximization criteria
3. Minimizing pressure drop losses
4. Corrosion and fouling
5. Structural design to satisfy fabrication and operational requirement

Interfacial energy balances require that the TE device heat flow at the hot-side and cold-sides (Equations (3.4a) and (3.4b)) must equal the heat exchanger heat flows dictated by Equations (3.12) and (3.13). Since it is desirable to match up the interfacial contact area between the TE device and heat exchanger, this furthermore requires that interfacial heat fluxes between the TE device and heat exchangers must also match.

Recent studies have also shown that in the case of maximizing TE system power, a critical thermal design criterion also derives from the key relationship:

$$\left(\frac{P}{P_{max}}\right) = f\left[\left(\frac{\sum_i R_{h,th,i}}{\sum_i R_{c,th,i}}\right), TE \cdot Material \cdot Properties\right] \qquad (3.42)$$

Through this relationship there exists an optimum ratio

$$\left(\sum_i R_{h,th,i} \bigg/ \sum_i R_{c,th,i}\right)_{opt}$$

that maximizes the power output; this is an additional criterion above and beyond normal TE design optimization conditions. These are the same hot- and cold-side thermal resistance summations as given in Equation (3.35). Recent research shows this relationship dictates that hot- and cold-side microtechnology heat exchangers must satisfy the critical criterion

$$\left(\frac{\sum_i R_{h,th,i}}{\sum_i R_{c,th,i}}\right) > 10 \rightarrow 30 \qquad (3.43)$$

to maximize TE system power output.

MHEX designs must also minimize parasitic pressure drop losses, as discussed in Section 3.3.1, which create TE system cold-side pumping power requirements, thereby lowering the net TE system power output, and limit potential heat transfer in energy recovery heat exchangers on the TE system hot side. This places important design constraints on possible microchannel dimensions and inlet/outlet fluid mani-folding approaches and designs in any given application.

Corrosion and fouling within the microchannel heat exchangers, which can severely limit TE system lifetimes and reliability, are a constant design concern, as in all heat exchanger designs. It is necessary to minimize these two effects by using appropriate flow filtration systems and surface coatings on the microchannel surfaces. An entire book chapter could be written on this topic alone and the reader is referred to various flow filter hydraulics and material science references for further information on this topic.

Microtechnology heat exchangers in TE systems must satisfy crucial structural stress and displacement requirements in their operational environments—including structural stresses induced by compression forces, thermal expansion induced forces, and vibrational and shock forces—just as in other conventional heat exchange applications. MHEX designs have additional structural requirements during their fabrication processes because of their typically small fin, channel, and interlayer thickness dimensions that are often on the order of only a few millimeters. Buckling stresses and other deformation stresses in improperly designed devices can lead to channel deformations, yielding of critical heat transfer surfaces, and delamination or voids at critical interfaces during fabrication processes.

3.3.2.2 Design Methodologies

Hendricks [26] and Krishnan, Leith, and Hendricks [30] have discussed various design approaches and methodologies to microchannel heat exchanger design for different configurations. These two references discuss methodologies and thermal transfer and fluid dynamic relationships to determine the UA and thermal resistance, R_{th}, of different MHEX designs (see Equations (3.35) and (3.36)). It is critical to determine the impacts on the MHEX UA of varying channel dimensions, flow rates, and parasitic pressure drops, which translate into pumping power. Figure 3.13 shows and exemplifies the typical UA-pumping power relationship or "map" as these design parameters vary for a constant mass flow rate and heat exchanger footprint area in a water–copper MHEX design similar to that shown in Figure 3.12. The key constant

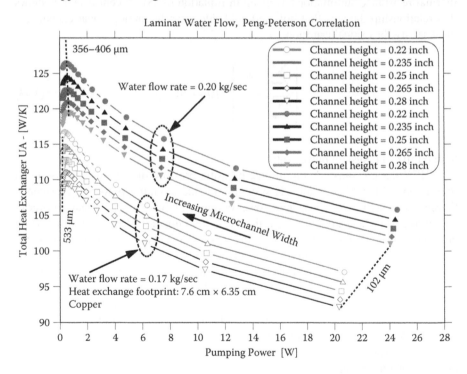

FIGURE 3.13 Microchannel heat exchanger UA versus pumping power in various water-copper microchannel designs. (Hendricks, T. J. 2008. *Proceedings of American Society of Mechanical Engineers 2nd International Conference on Energy Sustainability,* Jacksonville, FL, paper no. ES2008-54244. Used with permission from American Society of Mechanical Engineers, New York.)

MHEX footprint condition in this design analysis is often imposed by heat flux and TE device interface requirements in a given TE system application. Figure 3.13 clearly shows a typical relationship of UA with pumping power as the microchannel widths decrease across a wide range of potential design parameters for a given frontal area and heat exchanger footprint area. UA generally increases sharply to a maximum for small pressure drop and pumping power increases as microchannel widths decrease from the 503 to 350–400 μm point. This desirable optimum design regime is crucial to achieving high-performance designs for TE power systems.

The maximum UA point as pumping power varies is created by a complex interplay between total heat transfer area, convective heat transfer coefficient, and fluid pressure drops for the constant mass flow and heat exchanger footprint conditions [26]. As microchannel widths decrease further below the 350–400 μm range, UA decreases with the further penalty of larger pressure drops and pumping powers, creating an undesirable design regime that should be avoided. This UA maximum is a unique design point that must be determined for any particular design application using either gas or liquid flow. Increasing mass flow rates generally increases UA in this analysis as microchannel velocities increase, thereby increasing convective heat transfer coefficients, while increasing microchannel heights actually can decrease UA as microchannel velocities decrease, fin efficiencies decrease, and hydraulic diameters increase—all effects that decrease the convective heat transfer coefficient. This UA–pumping power relationship or "map" and associated optimum design regions are significantly influenced by the convective heat transfer correlation used in the analysis and the assumptions and constraints used (i.e., constant mass flow rate, constant pressure drop). However, Figure 3.13 generally shows the methodology and approach used in TE system design integration. It is paramount to quantify this UA–pumping power map across the entire possible design domain and uniquely determine optimum MHEX designs and off-optimum design sensitivities to critical design parameters for any given TE energy recovery system application.

Interface heat flux conditions at the heat exchanger–TE device interface are equally important in TE system designs. A common design goal is to maximize this interface heat flux—and therefore the MHEX heat flux—in order to maximize TE system power output, while simultaneously satisfying TE device footprint requirements created and driven by TE device design optimization criteria. Figure 3.14(a) exemplifies a typical relationship between interface heat flux and microchannel dimensions across a wide spectrum of design parameters for constant frontal area and heat exchanger footprint area (noted in the figure) in a water–copper MHEX design similar to that shown in Figure 3.12. Interfacial heat flux increases to maximum levels at about 350–400 μm for various microchannel heights in Figure 3.14(a); this condition corresponds to the maximum UA condition and desired design regime in Figure 3.13. This result again demonstrates that there can be optimum microchannel dimensions in any given application and the exact point of heat flux maximum varies slightly with microchannel height along the loci of maximums (dashed line) depicted in Figure 3.14(a). Interfacial heat flux decreases significantly as microchannel widths continue decreasing below 350 μm due to decreasing UA (in Figure 3.13) and subsequent heat transfer for the constant mass flow and frontal area conditions. The effect of water flow rate on interfacial heat flux, also shown in Figure 3.14(a),

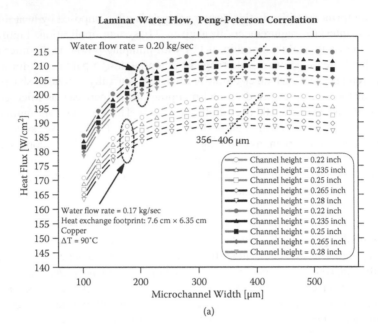

(a)

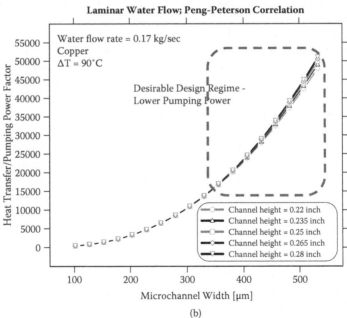

(b)

FIGURE 3.14   (a) Heat flux and (b) heat transfer/pumping power factor in various water–copper microchannel designs. (Hendricks, T. J. 2008. *Proceedings of American Society of Mechanical Engineers 2nd International Conference on Energy Sustainability,* Jacksonville, FL, paper no. ES2008-54244. Used with permission from American Society of Mechanical Engineers, New York.)

demonstrates that optimum microchannel dimensions are not changing as mass flow rate varies.

Figure 3.14(b) shows the effective heat transfer rate to required pumping power ratio, Γ [26], for the same range of microchannel widths and heat exchanger frontal area and footprint assumptions in Figure 3.14(a). It is clear that Γ is increasing sharply in design regions associated with maximum interfacial heat flux and continues increasing sharply at larger microchannel widths, due to smaller pumping powers with simultaneously small decreases in the interfacial heat flux in Figure 3.14(a). This is a highly desirable design regime as the loss in interfacial heat flux can be insignificant as Γ continues increasing sharply. Similarly to Figure 3.13 results, the associated optimum heat flux design regions are significantly influenced by the convective heat transfer correlation used in the analysis and the assumptions and constraints used. Nevertheless, Figure 3.14 and Hendricks [26] show the methodology and approach to microchannel heat exchanger design for TE energy recovery system applications. It is paramount to quantify the heat flux–Γ relationships and design regimes, as well as sensitivities and trade-offs across the entire possible design domain, and uniquely determine optimum heat flux designs and off-optimum design sensitivities to critical design parameters in any given TE system application.

Figure 3.15 shows similar UA–pumping power relationships and heat flux–Γ relationships for a helium gas microchannel heat exchanger design for a constant heat exchanger frontal area and footprint area. Many of the same trends and behavior are exhibited in this gas flow design, as shown in Figures 3.13 and 3.14, except that pumping powers are exorbitantly high and Γ values are therefore quite low in this case for the heat exchanger frontal area and footprint area given. The same, more desirable heat flux–Γ design regimes exist at large microchannel widths. These generally higher pumping powers for gas flow microchannel heat exchanger designs create serious design challenges in TE energy recovery systems, which still warrant further research and engineering work.

Krishnan, Leith, and Hendricks [30] discussed analysis methodologies and design techniques for employing and optimizing microchannel honeycomb designs to enhance gas flow heat transfer in TE systems. This work specifically focused on microchannel designs for automotive heating and air conditioning system applications.

3.3.2.3   Examples of TE/Microtechnology Integration

There are recent examples [10,11,25,26,30–32] where microtechnology heat exchange technology was investigated or implemented in TE systems. Figure 3.16 displays a recent rectangular honeycomb copper microchannel design that was demonstrated and characterized for a TE cooling system in an automotive application. This particular microchannel heat exchanger design occupied 8 cm × 4 cm × 2 cm in volume and demonstrated low pressure drops (0.52 kPa at ~18 cubic feet per minute air flow) and low thermal resistances near 0.1 C/W, and it was capable of 250–300 W of thermal transfer at temperature and flow rate conditions consistent with automotive heating and air conditioning requirements [30]. Wang et al. [32] demonstrated successful performance of several microtechnology heat exchangers incorporated within a unique, pioneering hybrid TE–organic Rankine cycle system, which integrated a TE topping cycle with an organic Rankine cycle and vapor

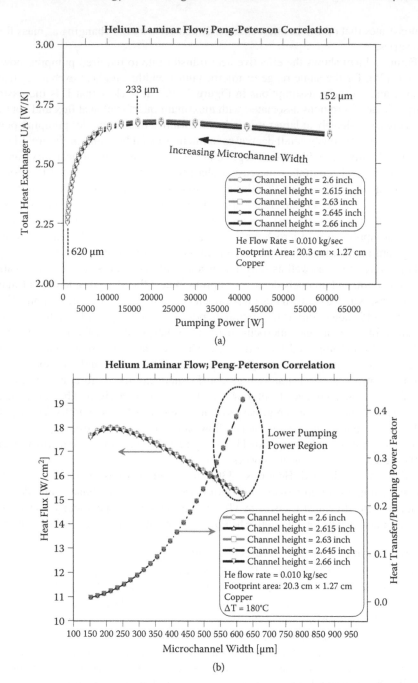

FIGURE 3.15　(a) Heat exchanger UA and (b) hot-side heat flux in helium flow microchannel designs. (Hendricks, T. J. 2008. *Proceedings of American Society of Mechanical Engineers 2nd International Conference on Energy Sustainability,* Jacksonville, FL, paper no. ES2008-54244. (Used with permission from American Society of Mechanical Engineers, New York.)

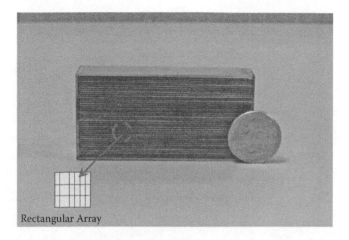

FIGURE 3.16 Recent rectangular honeycomb microchannel design for TE Systems in automotive application.

compression system to provide simultaneous power production and cooling capacity from a waste-heat-driven source. The TE subsystem used an air microchannel heat exchanger on the hot side and a liquid R-245fa microchannel heat exchanger on its cold side. The organic Rankine cycle operating with R-245fa working fluid used a microchannel heat exchanger for its boiler and recuperator devices [32]. All these microchannel heat exchangers operated stably with low pressure drops and good thermal performance per their design specifications; the organic Rankine cycle (ORC) boiler and recuperator microchannel heat exchangers exhibited actual thermal effectiveness generally above 90%.

There is a common misconception that microchannel heat exchangers necessarily produce high pressure drops and, therefore, have high pumping power. This is not true, as evidenced by the designs in Figure 3.16 and Wang et al. [32]. In fact, if a microchannel heat exchanger is properly designed and properly scaled, its pressure drop need not be any higher than a conventional macrochannel heat exchanger design for a given heat transfer duty and mass flow rate. This also has been shown in numerous examples in Yang and Holladay's work [33]. The cost of microchannel systems also is being driven down as production volumes increase in potential high-volume applications for automotive and heavy vehicles; industrial processes; residential and commercial heating, ventilation, and air conditioning systems; and power generation.

3.4 STRUCTURAL DESIGN AND CONSIDERATIONS IN THERMOELECTRIC SYSTEMS

The TE modules in TE heat recovery systems are generally constructed of multiple TE elements that are often bulk elements arranged in an array that is thermally in parallel and electrically in series, as shown in Figure 3.17, or as thin-film elements that encounter a substantial temperature differential across a length dimension

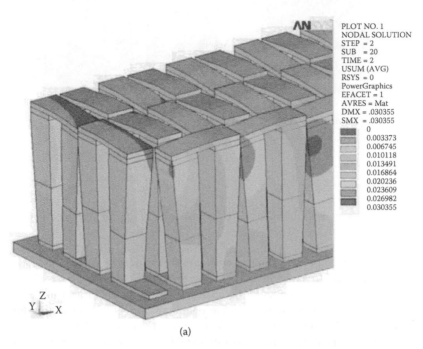

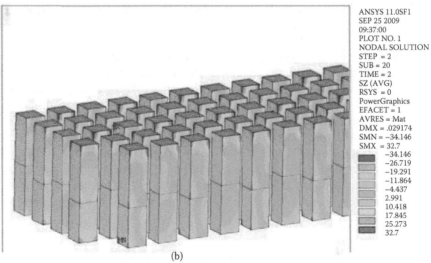

FIGURE 3.17 Top (a): magnified structural displacements; bottom (b): resulting stresses in a TE module from compressive and expansive forces (displacements in millimeters; stresses in megapascals). Note: The TE hot side is on the top of each element shown.

perpendicular to and much greater than the thinnest film dimension. The structural design and analysis is a critical aspect of high-temperature thermoelectric heat recovery systems because of thermally induced expansion stresses from materials with even slightly mismatched coefficients of thermal expansion, compression stresses, and tensile stresses developed within the TE elements and module. In bulk element arrays, the TE module is commonly under significant compression of 30 psi up to 200 psi in order to create adequate thermal transport across critical hot- and cold-side interfaces between heat exchanger and TE device surfaces. This compression can be applied at room temperature or the system can be designed to achieve these compressive pressures as it expands to higher temperatures, depending on the details of the design. In thin-film element designs the thermal contact can be supplied compressively also, but often compression is exerted perpendicular to the element length at the element ends in some fashion that is highly design dependent. In any event, the hot side of the TE element can be at temperatures of 600 to 1300 K, while the cold side can be held at 350 to 500 K.

In addition, the TE elements themselves are electrically connected with a variety of electrically conducting and diffusion barrier materials, which often have different coefficients of thermal expansion from those of the base TE materials. In any event, the combination of compression and thermally induced expansions as the system heats up creates very complex structural displacements and stresses in the TE elements and generates the potential for large structural stresses that can damage the TE element and modules. Figure 3.17 shows a magnified structural displacement map created from a typical structural analysis using the ANSYS® software package for a typical rectangular parallelepiped TE element/module design. It highlights how the elements respond to the expansive and compressive forces encountered in the module and in a system. It is not only the TE elements that expand, but the electrical connecting materials and diffusion barrier materials also expand, putting complex tilting and distortion forces on the elements.

As one can see, the TE elements can experience complex expansive displacements and tilting displacements that create complex structural tensile and compressive stresses at various locations in the TE device. Element corners at the rectangular parallelepiped TE device hot side are particularly susceptible to high tensile stresses (i.e., dark regions in the top corner areas), for example, as the device components expand during heating to operational conditions. This fact does not necessarily change whether one is considering a bulk material TE element/module design as shown in Figure 3.17 or a thin-film TE element/module design that others have considered in some TE heat recovery applications.

Figure 3.17 exemplifies the type of structural analysis that is required in designing TE elements and modules for thermoelectric heat recovery systems. These element and module structural analyses are generally quite complex and three dimensional in nature. They require strict and comprehensive evaluation of boundary conditions and often require multiple "load steps" as boundary conditions change during fabrication and operation to analyze the TE element, component, and module stresses and displacements properly for all expected environments. The required material properties for performing these analyses are typically Young's modulus, Poisson's ratio, coefficient of thermal expansion, and mechanical fracture strengths for each

TABLE 3.4

Typical Structural Properties of Common TE Materials Compared to Semiconductors and Common Metals [34, 35]

Material	Young's Modulus, E (GPa)	Poisson's Ratio, υ	Coefficient of Thermal Expansion (/°C)	Fracture Toughness, K_c (MPa–m$^{0.5}$)	Fracture Strength, σ_f/(MPa)
Si	163	0.22	2.6×10^{-6}	0.7	247
Ge	128	0.21	5.9×10^{-6}	0.60	231–392
GaAs	117	0.24	6.9×10^{-6}	0.46	66
PbTe	58	0.26	19.8–20.4×10^{-6}		
LAST/-T (Michigan State Univ.)	24.6–71.2	0.24–0.28		—	15.3–51.6
LAST (Tellurex Corp.)	54–55	0.27–0.28	21×10^{-6}	—	15–38 (ROR)
LASTT (Tellurex Corp.)	46.3–46.8	0.26–0.27	21×10^{-6}	—	25–40 (ROR); 46–58 (BOR)
ZnSe	76.1	0.29	8.5×10^{-6} (293–573 K)	0.9	~60
Zn$_4$Sb$_3$	57.9–76.3			0.64–1.49	56.5–83.4
Bi$_2$Te$_3$	40.4–46.8	0.21–0.37	14.4×10^{-6} ($\perp$); 21×10^{-6} ($\|\|$)		8–166
Skutterudites					
Ba$_{0.05}$Yb$_{0.2}$Co$_4$Sb$_{12}$ (n)	136 (n)	0.14–0.25 (n)	12.2×10^{-6} (n)	1.7 (n)	86 (n)
Ce$_{0.85}$Fe$_{3.5}$Co$_{0.5}$Sb$_{12}$ (p)	133 (p)	0.22–0.29 (p)	14.5×10^{-6} (p)	1.1–2.8 (p)	37 (p)
Stainless steel	205		18.5×10^{-6}		300
Copper	129		15–18×10^{-6}		198

Source: Hall, B. D. et al. 2007. Technical presentation at Materials Science and Technology 2007 Conference, Detroit, MI.

Notes: $\perp$ = perpendicular to current flow (element length); $\|\|$ = parallel to current flow (element length); ROR = ring-on-ring fracture strength; BOR = ball-on-ring fracture strength; LAST = lead–antimony–silver–telluride; LASTT and LAST/-T: lead–antimony–silver–telluride–tin.

of the materials used for each component in the TE module design. Table 3.4 shows some typical structural property values of common TE materials compared to semiconductors and common metals.

There are generally no consistent "rules of thumb" or set of equations that one can employ even to get close to accurate structural analysis results. What is generally required is a complete three-dimensional analysis using a general structural analysis code, such as ANSYS®, COMSOL®, or other such structural analysis software packages. Furthermore, the TE element and module structural analyses must be tightly coupled with the TE optimization and performance analyses discussed earlier in this chapter to develop optimum, survivable TE element and module designs that satisfy all operational requirements in the environments anticipated for any thermoelectric heat recovery system. For example, in structural analyses exemplified in Figure 3.17, it is often found that longer, thinner TE elements lower the destructive tensile stresses that develop. However, the TE design optimization equations shown in Section 3.2 dictate that longer TE elements will necessarily produce lower power designs. Consequently, requirements for maximizing power output from the TE heat recovery system are in conflict with the requirement to lower structural stresses and ensure TE element and module survival during operational conditions and environments. Therefore, it is imperative that the structural design and analysis be strongly integrated with the thermoelectric and thermal design optimization processes described previously in achieving a truly optimum design for TE heat recovery systems.

3.5 CONCLUSION

Advanced, high-performance TE energy recovery systems will have a unique and critical role in recovering waste energy in automotive and industrial applications worldwide, with the subsequent benefit of helping to increase global energy efficiency. Optimal design of these systems requires design optimization accounting for three crucial, interdependent design areas: thermoelectric design, thermal design, and structural design. Copious consideration and attention to detail in these three design regimes are equally important to that of obtaining high-performance TE materials in developing TE devices and systems that achieve their full performance potential for these applications. This chapter provides the foundation and methodology for addressing these three critical system design facets and their role in achieving high-performance TE waste energy recovery systems and solutions. If one wants this technology to achieve its full performance potential in future applications, then one must explore comprehensive system design domains using the techniques discussed herein.

ACKNOWLEDGMENTS

The authors acknowledge and thank Professor Eldon Case and his group at the Department of Chemical Engineering and Material Science, Michigan State University, East Lansing, Michigan, for their mechanical property contributions in Table 3.4. The authors also thank Mr. Naveen Karri, Engineering Mechanics and Structural Materials Group, Radiological and Nuclear Science and Technology Division, Pacific Northwest National Laboratory, for his support and expertise in preparing critical figures in this chapter. The authors would also like to thank Virginia M. Sliman and Brenda L. Langley at the Pacific Northwest National Laboratory for their rigorous technical editing of this manuscript and their great editorial recommendations.

REFERENCES

1. *Transportation energy data book,* ed. 29. 2010. US Department of Energy, Office of Energy Efficiency and Renewable Energy, Vehicles Technology Program. ORNL-6985, Oak Ridge National Laboratory, Oak Ridge, TN. http://cta.ornl.gov/data/index.shtml

2. US Department of Energy, Industrial Technologies Program, Energetics, Inc. & E3M, Inc. 2004. Energy use, loss and opportunities analysis: US Manufacturing and Mining.

3. Hendricks, T. J., and W. T. Choate. 2006. Engineering scoping study of thermoelectric generator packages for industrial waste heat recovery. US Department of Energy, Industrial Technology Program, http://www.eere.energy.gov/industry/imf/analysis.html

4. Hendricks, T. J., and J. A. Lustbader. 2002. Advanced thermoelectric power system investigations for light-duty and heavy-duty vehicle applications: Part I. *Proceedings of the 21st International Conference on Thermoelectrics,* Long Beach, CA, IEEE catalogue no. 02TH8657, 381–386.

5. Hendricks, T. J. and J. A. Lustbader. 2002. Advanced thermoelectric power system investigations for light-duty and heavy-duty vehicle applications: Part II. *Proceedings of the 21st International Conference on Thermoelectrics,* Long Beach, CA, IEEE Catalogue no. 02TH8657, 387–394.

6. Crane, D. T., J. W. LaGrandeur, and L. E. Bell. 2010. Progress report on BSST led, U.S. DOE automotive waste heat recovery program. *Journal of Electronic Materials* 39 (9): 2142–2148.

7. Angrist, S. W. 1982, *Direct energy conversion,* 4th ed. Boston, MA: Allyn and Bacon.

8. Rowe, D. M., ed. 1995. *CRC handbook of thermoelectrics.* Boca Raton, FL: CRC Press.

9. Hogan, T. P., and T. Shih. 2005. Modeling and characterization of power generation modules based on bulk materials. In *Thermoelectrics handbook: Micro to nano,* ed. D. M. Rowe. Boca Raton, FL: CRC Press.

10. Hendricks, T. J., N. K. Karri, T. P. Hogan, and C. J. Cauchy. 2010. New thermoelectric materials and new system-level perspectives using battlefield heat sources for battery recharging. *Proceedings of the 44th Power Sources Conference,* Institute of Electrical and Electronic Engineers Power Sources Publication, technical paper no. 28.2, pp. 609–612.

11. Crane, D. T. 2011. An introduction to system level steady-state and transient modeling and optimization of high power density thermoelectric generator devices made of segmented thermoelectric elements. *Journal of Electronic Materials* 40: 561–569.

12. Cobble, M. H. 1995. Calculations of generator performance. *CRC handbook of thermoelectrics,* Boca Raton, FL: CRC Press.

13. Swanson, B. W., E. V. Somers, and R. R. Heikes. 1961. Optimization of a sandwiched TE device. *Journal of Heat Transfer* 83: 77–82.

14. Crane, D. T. 2003. Optimizing thermoelectric waste heat recovery from an automotive cooling system. PhD dissertation, University of Maryland, College Park.

15. Crane, D. T., and L. E. Bell. 2006. Progress towards maximizing the performance of a thermoelectric power generator. *Proceedings of the 25th International Conference on Thermoelectrics,* Vienna, Austria: IEEE, pp. 11–16.

16. Crane, D. T., and L. E. Bell. 2009. Design to maximize performance of a thermoelectric power generator with a dynamic thermal power source. *Journal of Energy Resources Technology* 131: 012401-1–012401-8.

17. Crane, D. T., and G. S. Jackson. 2004. Optimization of cross flow heat exchangers for thermoelectric waste heat recovery. *International Journal of Energy Conversion and Management* 45 (9–10): 1565–1582.

18. Sherman, B., R. R. Heikes, and R. W. Ure, Jr. 1960. Calculations of efficiency of TE devices. *Journal of Applied Physics* 31 (1): 1–16.

19. Domenicali, C. A. 1953, Irreversible thermodynamics of TE effects in inhomogeneous, anisotropic media. *Physics Reviews* 92 (4): 877–881.
20. Mahan, G. D. 1991. Inhomogeneous TEs. *Journal of Applied Physics* 70 (8): 4551.
21. Skrabek, E. A., and D. S. Trimmer. 1995. Properties of the general TAGS system. In *CRC handbook of thermoelectrics*, ed. D. M. Rowe, 267–275. Boca Raton, FL: CRC Press LLC.
22. Dughaish, Z. H. 2002. Lead telluride as a thermoelectric material for thermoelectric power generation. *Physica B* 322: 205–223.
23. Antonova, E. E., and D. C. Looman. 2005. Finite elements for thermoelectric device analysis in ANSYS®. *Proceedings of 2005 24th International Conference on Thermoelectrics*, pp. 215–218.
24. Kays, W. M., and A. L. London. 1984. *Compact heat exchangers*, 3rd ed. New York: McGraw–Hill.
25. Hendricks, T. J., and N. K. Karri. 2009. Micro- and nano-technology: A critical design key in advanced thermoelectric cooling systems. *Journal of Electronic Materials* 38 (7): 1257–1267, DOI: 10.1007/s11664-009-0709-3, Springer Publishing, New York.
26. Hendricks, T. J. 2008. Microtechnology—A key to system miniaturization in advanced energy recovery and conversion systems. *Proceedings of American Society of Mechanical Engineers 2nd International Conference on Energy Sustainability*, Jacksonville, FL, paper no. ES2008-54244.
27. Hendricks, T. J. 2006, Microchannel & minichannel heat exchangers in advanced energy recovery & conversion systems. *Proceedings of the ASME 2006 International Mechanical Engineering Congress and Exposition*, IMECE2006—Advanced Energy System Division. New York: American Society of Mechanical Engineers, paper no. IMECE2006-14594.
28. Incropera, F. P., and D. P. Dewitt. 1990. *Fundamentals of heat and mass transfer*, 3rd ed. New York: John Wiley & Sons.
29. Liao, S. M., and T. S. Zhao. 2002. Measurements of heat transfer coefficients from supercritical carbon dioxide flowing in horizontal mini/micro channels. *Journal of Heat Transfer, Transactions of the ASME* 124:413–420.
30. Krishnan, S., S. Leith, and T. J. Hendricks. 2012. Enhanced gas-side heat transfer in rectangular micro-honeycombs. *Proceedings of the ASME 2012 6th International Conference on Energy Sustainability & 10th Fuel Cell Science, Engineering and Technology Conference*. New York: American Society of Mechanical Engineers, paper no. ESFuelCell2012-91223.
31. Hendricks, T. J., T. P. Hogan, E. D. Case, C. J. Cauchy, N. K. Karri, J. D'Angelo, C.-I. Wu, A. Q. Morrison, and F. Ren. 2009. Advanced soldier-based thermoelectric power systems using battlefield heat sources. In *Energy harvesting—From fundamentals to devices*, ed. H. Radousky, J. Holbery, L. Lewis, and F. Schmidt (Materials Research Society Symposium Proceedings 1218E, Warrendale, PA, 2010). Paper ID no. 1218-Z07-02. (Proceedings of the Materials Research Society 2009 Fall Meeting, Symposium Z, paper ID no. 1218-Z07-02, Boston, MA, 2009.)
32. Wang, H., T. J. Hendricks, S. Krishnan, and R. Peterson, 2011. Experimental verification of thermally activated power and cooling system using hybrid thermoelectric, organic Rankine cycle and vapor compression cycle. *Proceedings of the 9th Annual International Energy Conversion Engineering Conference* (Reston, VA: American Institute of Aeronautics and Astronautics, Inc.), paper no. AIAA 2011-5983.
33. Wang, Y., and J. D. Holladay. 2005. *Microreactor technology and process intensification*. ACS Symposium Series 914, American Chemical Society, Oxford University Press.

34. Hall, B. D., J. L. Micklash, J. .R Johnson, T. P. Hogan, and E. D. Case. 2007. A review of mechanical properties for thermoelectric materials. Technical presentation at Materials Science and Technology 2007 Conference, Detroit, MI.
35. Morrison, A. Q., E. D. Case, F. Ren, A. J. Baumann, T. J. Hendricks, C. Cauchy, and J. Barnard. 2012. Elastic modulus and bioxial fracture strength of thermally fatigued hot pressed LAST and LASTT thermoelectric matierals. *Materials Chemistry and Physics* 134: 973–987.

4 Thermopower Wave-Based Micro- and Nanoscale Energy Sources

Sumeet Walia and Kourosh Kalantar-Zadeh

CONTENTS

4.1 INTRODUCTION

There is a growing need to reduce the size of current energy sources in order to realize fully the potential of many fascinating systems, such as "smart dust," integrated microelectromechanical (MEMS) systems and biomedical electronic devices such as cardiac microdefibrillators [1]. Large-scale deployment of applications such as hybrid cars with high efficiency has also been hampered for a long time due to lack of power sources that match their dimensional scale and deliver the required power densities [2].

Miniaturization of energy sources is also highly desirable for applications such as heat engines, where a large power surge is required solely at the start. However, heavy and bulky batteries have to be accommodated for the entire journey, thereby reducing the efficiency of the entire system.

Batteries have been the most commonly used power source for a long time. Although there have been recent advances reported in fabricating three-dimensional Li ion micro- and nanobatteries, their specific power and energy densities are still quite low (0.3 kW/kg and 0.4 kWh/kg, respectively) [2]. Additionally, a reduction in size makes them lose their energy discharge capabilities. The reduction in size is also hampered by disruption of ionic flow near the electrodes and by electrical shorts that impede their performance [3]. Consequently, batteries continue to be heavy and bulky, thereby limiting their use for micro- and nanoscale applications.

Fuel cells are another commonly used energy generation technology. However, they require large volumes and expensive electronics while exhibiting very low specific power (power-to-mass) ratios [4,5].

Energy can also be stored and released at high rates using supercapacitors. Although such systems exhibit a high specific power, they are rendered inefficient due to their high self-discharge rate. They have to be charged at regular intervals in order to maintain the stored energy [6].

In addition to the aforementioned limitations, all such energy sources are capable of generating only DC output; however, for practical applications, an AC output is highly desirable. Thus, none of the currently used energy generation technologies have so far enabled reliable miniaturized energy sources with a high specific power.

Consequently, novel techniques for energy generation need to be explored in order to achieve true miniaturization of power sources. New thermopower, wave-based energy sources are particularly attractive as they are capable of generating higher specific power (power-to-mass ratio) compared to conventional sources such as batteries and fuel cells, especially at smaller dimensions.

Choi et al. first demonstrated that chemically driven carbon nanotube-guided thermopower waves are capable of producing specific powers as large as 7 kW/kg [7]. In their work, aligned arrays of multiwalled carbon nanotubes coated with a several-nanometers-thick layer of cyclotrimethylene-trinitramine (TNA) were developed (Figure 4.1). In such devices, the energy is stored in the chemical bonds of the fuel (TNA, in this case), which enables them to maintain their energy for a long period of time. A laser or microheater was used at one end to initiate an exothermic decomposition reaction of TNA. The thermally conductive MWNT conduit

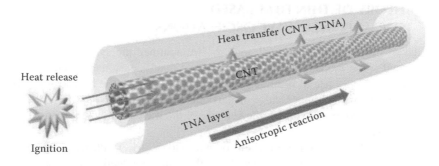

FIGURE 4.1 Schematic of the thermopower wave system showing the reaction propagation mechanism across a TNA-coated carbon nanotube.

accelerated the reaction along their length, creating a unidirectional thermal wave up to 10,000 times faster than the reaction wave in bulk. This thermal wave also entrains electronic carriers, resulting in output power, and, hence, these waves were coined as "thermopower waves."

The main limitation of the MWNT-based thermopower systems is the low voltage generated, which is generally in the range of 30–50 mV with a maximum reported voltage of 200 mV for masses in the range of just a few milligrams. Additionally, the amplitude of the oscillations is also low (20–30 mV generally). The reason for this limitation is that MWNTs exhibit a relatively low Seebeck coefficient (80 µV/K). In order to increase the output voltage and thus improve the performance of such systems, materials such as bismuth telluride (Bi_2Te_3), antimony telluride (Sb_2Te_3), and transition metal oxides such as zinc oxide (ZnO) with large respective absolute Seebeck coefficient values of 287, 243, and 367 µV/K can potentially replace MWNTs as the core thermoelectric materials for thermopower energy generation systems. Another desirable property of the core thermoelectric material is a high electrical conductivity, which results in a larger power output.

This chapter provides an overview of the thermopower wave technology. Such systems, based on MWNTs and thin films of thermoelectric materials such as Bi_2Te_3, Sb_2Te_3, and ZnO, will be discussed. These materials exhibit a high Seebeck coefficient and high electrical conductivities. A high thermal conductivity is also required for sustaining the propagation of the thermopower waves. As the thermal conductivity of Bi_2Te_3 and Sb_2Te_3 materials is low, thermally conductive alumina (Al_2O_3) is used as the substrate in order to compensate for this deficiency. Additionally, in order to show the influence of the substrate thermal conductivity on the behavior of the thermopower waves, a substrate with low thermal conductivity (terracotta) is also used.

It has been shown that Bi_2Te_3- and Sb_2Te_3-based thermopower systems are capable of generating oscillatory voltages up to 400 mV with complementary polarities and specific power as high as 1 kW/kg. ZnO-based thermopower devices generated the highest voltage and oscillations of the order of 500 mV and a specific power of up to 0.5 kW/kg [2,8,9].

4.2 THEORY OF THIN FILM-BASED THERMOPOWER WAVE OSCILLATIONS

In order to simplify the theoretical analysis, we assumed the system to be a one-dimensional space model and transformed it to a nondimensional system in order to predict the occurrence of oscillatory combustion linked to the heat losses in the system. The theoretical approach is of particular importance as it can be used to predict the behavior of the system, the velocity of the wave front, and the possibility of oscillations.

The thermoelectric film is thin when compared to the Al_2O_3 substrate. Therefore, the majority of the heat is exchanged with the thermally conductive Al_2O_3 substrate. As a result, we consider one-dimensional approximations for both the fuel/Al_2O_3 system and the fuel/terracotta system in our calculations. It means that fuel exchanges heat either with a thermally conductive substrate of alumina with $k = 20$ W/m.K or the much less thermally conductive substrate terracotta with $k = 1$ W/m.K. Assuming the solid fuel combustion to be exothermic and oxygen supplied in excess and governed by Arrhenius kinetics with volumetric heat loss, the dimensional equations for the conservation of heat and mass are [10]

$$\rho c_p \frac{\partial T}{\partial t} = k \frac{\partial^2 T}{\partial x^2} + \rho Q A w e^{-E/RT} - \frac{hS}{V}(T - T_a) \tag{4.1}$$

$$\rho \frac{\partial w}{\partial t} = -\rho A w e^{-E/RT} \tag{4.2}$$

where
T and w denote the temperature and the concentration of the fuel, respectively
x and t describe space and time variables
ρ is the density of the fuel (kg/m³)
c_p is the specific heat of the fuel (J/kg.K)
k is the thermal conductivity of the fuel (J/s.m.K)
Q is the heat of reaction (J/kg)
A is the pre-exponential rate constant (/s)
E is the activation energy (J/mol)
R is the universal gas constant (8.314 J/mol.K)

Heat is transferred from the fuel layer to the Bi_2Te_3/substrate layer and to the surroundings. As we are using a one-dimensional averaged model, the term

$$\frac{hS}{V}(T - T_a) \tag{4.3}$$

models both the transfer to Bi_2Te_3/substrate and the Newtonian cooling to the ambient surrounding, which is at a temperature T_a [10]. In the latter case, S/V is the surface-area-to-volume ratio of the fuel (/m) and h is the heat transfer coefficient from the fuel to the quiescent surroundings (J/s.m^2.K) [10], which is typically quite small. The term $\rho QAwe^{-E/RT}$ in Equation (4.1) accounts for the exponential decomposition reaction of the fuel.

As a first step toward analyzing the combustion, we applied the model presented by Mercer, Weber, and Sidhu [10]. Defining the nondimensional temperature to be $u = RT/E$ and rescaling space and time coordinates by

$$\sqrt{\frac{\rho QAR}{kE}} \text{ and } \frac{QAR}{c_p E},$$

respectively, leads to the nondimensional version of the governing equations:

$$\frac{\partial u}{\partial t} = \frac{\partial^2 u}{\partial x^2} + we^{-1/u} - \ell(u - u_a) \tag{4.4}$$

$$\frac{\partial w}{\partial t} = -\beta we^{-1/u} \tag{4.5}$$

where we define

$$\beta = \frac{Ec_p}{QR}, \tag{4.6}$$

a parameter related to the properties of the fuel, and the nondimensional volumetric heat transfer,

$$\ell = \frac{hSE}{VR\rho QA}. \tag{4.7}$$

Depending on the value of system parameters, we can predict thermopower waves with or without oscillations [11,12]. For simplification in our theoretical analysis, we start with a one-dimensional space model and transform it to a nondimensional system to predict the occurrence of oscillatory combustion linked to the heat losses.

The determination of all the necessary parameter values is the key to being able to apply the model to a real system. The solution of the model presented by Mercer et al. [10] assumes that the ambient temperature is absolute zero ($u_a = 0$). Ambient temperatures are typically very small compared to the reaction temperature and, hence, this assumption has little effect on the overall behavior of the solutions to this model.

For the fuel (which is a combination of nitrocellulose and sodium azide in our case), the value parameters are as follows: S/V = 1/240/μm, ρ = 1600 kg/m^3, c_p = 1596 J/kg.K, Q = 4.75 × 10^6 J/kg, E = 1.26 × 10^5 J/mol, A = 10^5 s^{-1}, and h = 2 × 10^{-3} J/s·m^2.K. In all cases, the β value is 5.09 for the nitrocellulose fuel layer. Two extreme conditions are now considered for the substrate:

a. *A highly thermally conductive substrate* (k = 20 W/m.K): This system is similar to the schematic shown in Figure 4.1. The fuel is deposited on Al_2O_3 (with a thin layer of Bi_2Te_3 in between) with a thermal conductivity of approximately 20 W.m^{-1}.K^{-1}. This results in the nondimensional heat loss parameter ℓ = 1.7 × 10−4. The oscillation period is directly related to β and ℓ. For β = 5.09, the critical heat loss value for the onset of oscillations is smaller than ℓ/β = 0.000035 (see reference 2 for details). Furthermore, the model predicts that there would be a time of approximately 0.05 s between peaks in the oscillating signal. In this case an oscillatory behavior is predicted; hence, the slower and faster propagation speeds are strongly dependent on very fine-scale details of the solution to the model; the slower one is of the order of 0.002 m/s and the faster one of an order of magnitude higher. Due to the oscillatory behavior, a two-dimensional model will better account for the various paths of heat conduction to provide accurate velocity predictions in this case.

b. *A substrate with a low thermal conductivity* (k = 1 W/m.K): The fuel is deposited on terracotta (with a thin layer of Bi_2Te_3 in between) with a thermal conductivity of approximately 1 W/m.K. The change in thermal conductivity of the substrate affects the heat loss parameter. In the case of terracotta, the heat loss parameter is obtained as ℓ = 2 × 10^{-5} and it is predicted that oscillations will cease to exist [2].

Next, we apply the model proposed by McIntosh, Weber, and Mercer [13] to predict the reaction propagation velocity for the fuel/terracotta system. The equation is as follows:

$$c - \frac{\ell}{c} = \frac{1}{c}e^{-\beta}\left(e^{2\ell\beta}/c^2 - e^{-\beta}\right) \tag{4.8}$$

where c is the wave speed in meters per second. Using Equation (4.3), the average reaction propagation velocity is estimated to be approximately 0.007 m/s. The velocity measured for experiments on the fuel/terracotta system is about 0.009 m/s, so the simulation agrees reasonably well with the measurements. Hence, the mathematical model is able to predict the oscillation period and the reaction propagation velocity reasonably well.

4.2.1 CHARACTERIZATION

4.2.1.1 X-Ray Diffraction (XRD)

XRD was carried out using a Bruker D8 DISCOVER microdiffractometer fitted with a GADDS (general area detector diffraction system). Data were collected at room temperature using CuKα-radiation (α = 1.54178 Å) with a potential of 40 kV and a current of 20 mA and filtered with a graphite monochromator in the parallel mode (175 mm collimator with 0.5 mm pinholes).

4.2.1.2 Scanning Electron Microscopy (SEM), Atomic Force Microscopy (AFM), and Raman Spectroscopy

Scanning electron microscopy (SEM) was conducted using the FEI Nova Nano SEM to investigate the structure of the ZnO films. Atomic force microscopy (AFM) was carried out on the ZnO films using an AFM-Bruker D3100 in order to assess their surface roughness. The Raman spectra were recorded with a system incorporating an Ocean Optics QE 6500 spectrometer and a 532 nm, 40 mW laser as the excitation source.

4.3 THERMOPOWER WAVE SYSTEMS

4.3.1 MULTIWALLED CARBON NANOTUBE (MWNT)-BASED THERMOPOWER SYSTEMS

Choi et al. first demonstrated the phenomenon of thermopower waves using MWNTs as a core thermoelectric material [14]. In their system, they synthesized MWNTs encased in a 7 nm thick annular coating of cyclotrimethylene trinitramine (TNA) (Figure 4.1).

The free-standing films exhibit high porosity (95%–99%) and are able to incorporate approximately 3–45 times TNA by mass. Subsequently, a NaN_3/water solution was applied and allowed to dry in ambient conditions.

The reaction was initiated at one end of the TNA-MWNTs using laser irradiation or a high-voltage electrical discharge. This resulted in a self-propagating reaction wave specifically along the direction of the nanotube orientation (Figure 4.1) at velocities more than four orders of magnitude larger than the bulk combustion rate of TNA.

Such a directional thermal wave evolves a corresponding thermopower wave in the same direction, resulting in a high specific power electrical pulse with a constant polarity. The measurement setup is shown in Figure 4.2(a). Ignition via laser irradiation at one end of the MWNT results in a voltage oscillation peak of the same duration as the corresponding reaction wave.

This demonstrates that the thermopower wave is uniquely distinct from conventional, static thermopower in the fact that single polarity pulses are observed over the duration of the reaction for the high-velocity waves. A moving thermal gradient across a conductor results in static regions of maximum, minimum, and zero voltages as a result of the Seebeck effect. However, the voltage signal obtained as a result of thermopower wave propagation is oscillatory in nature while exhibiting a constant polarity (Figure 4.2b); this suggests that the wave traverses the system faster than the cooling time of its posterior region, resulting in highly directional energy pulses [7,15].

As discussed earlier, the main limitation of MWNT-based thermopower wave systems was the low voltage and very limited oscillatory behavior. Therefore, in order to make such systems practically useful, the output from the thermopower wave-based devices needs to be enhanced. This prompted an investigation of other thermoelectric materials, such as Bi_2Te_3, Sb_2Te_3, and ZnO that exhibited high Seebeck coefficients and electrical conductivity. Thermopower devices based on

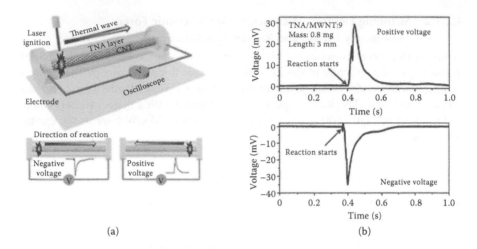

(a)   (b)

FIGURE 4.2   (a) Schematic of the experimental setup for a MWNT-based thermopower wave system and (b) output voltage profiles with opposite polarities obtained from MWNT-based thermopower wave system.

thin films of the previously mentioned thermoelectric materials have been developed and the results are discussed later in this chapter.

4.4   BI₂TE₃- AND SB₂TE₃-BASED THERMOPOWER WAVE SYSTEMS

In this section, thermopower wave generation is demonstrated in the thin film geometry by coupling exothermic chemical reaction of nitrocellulose to Bi_2Te_3 and Sb_2Te_3 films supported by a highly thermally conductive alumina (Al_2O_3) substrate in comparison to a much less thermally conductive terracotta (baked natural polysilicate) substrate (Figure 4.3) [2].

These devices produced a high specific power of the order of 1 kW/kg and generated voltages of the order of 100–150 mV with large oscillation amplitudes in the range of 50–140 mV. The generated voltage and the amplitude of oscillations for Bi_2Te_3-based devices were significantly larger than for MWNT-based thermopower devices.

Subsequently, Sb_2Te_3 was used as the complementary thermoelectric material. Sb_2Te_3 is a p-type semiconductor, while Bi_2Te_3 is an n-type [16,17]. Such a contrast ensures that the voltage output from these thermopower wave sources with identical characteristics (one based on Bi_2Te_3 and the other based on Sb_2Te_3) produce both positive and negative polarities. The combination of these two can result in an alternating signal, which is required for practical applications. Similarly to Bi_2Te_3, Al_2O_3 and terracotta are used as the substrates. They are chosen due to their contrasting thermal conductivities, which will help observe the effect of thermal properties of the system on the thermopower wave propagation. A detailed comparison of Sb_2Te_3- and Bi_2Te_3-based thermopower wave devices is presented later in this section.

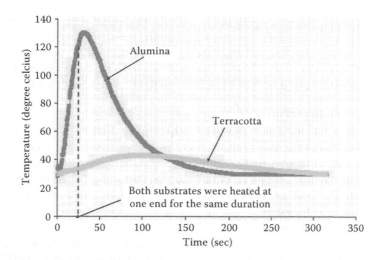

FIGURE 4.3 Difference in thermal conduction rates of Al_2O_3 and terracotta.

4.4.1 EXPERIMENTAL DETAILS OF THIN FILM THERMOPOWER WAVE SYSTEMS

4.4.1.1 Preparation of Fuel

The fuel used for the exothermic reaction was a combination of nitrocellulose ($C_6H_8(NO_2)_2O_5$) and sodium azide (NaN_3). Nitrocellulose was used due to its large enthalpy of reaction (4.75×10^6 J/kg). It was prepared by dissolving millipore nitrocellulose membranes (N8645, Sigma Aldrich) in acetonitrile (15 g/L). Using a pipette, this solution was then dropcast on the Bi_2Te_3 and Sb_2Te_3 films and left to dry. The acetonitrile evaporated, leaving an adhesive and solid nitrocellulose layer deposited on top of the thermoelectric materials. NaN_3 was used to serve as a primary igniter due to its low activation energy (40 kJ mol^{-1} for NaN_3, compared to 110-150 kJ mol^{-1} for nitrocellulose) [18,19]. NaN_3 (14314, Alfa Aesar) in aqueous solution (50 mg mL^{-1}) was then added on top of the nitrocellulose layer, using a pipette. The total thickness of the fuel layer was ~240 μm [2,8].

4.4.1.2 Fabrication of the Bi_2Te_3 and Sb_2Te_3 Films

The Sb_2Te_3 and Bi_2Te_3 films were deposited on thermally conductive Al_2O_3 substrates and nonthermal-conducting terracotta substrates using RF (radio frequency) magnetron sputtering, under identical deposition conditions. High purity (99%) Sb_2Te_3, Bi_2Te_3, and ZnO targets (Vin Karola Instruments) were used for sputtering. The chamber was pumped down to a pressure of 2×10^{-5} torr before starting the sputtering process.

Al_2O_3 substrates with linear dimensions of 12–15 mm (L) × 4–6 mm (W) × 100 μm (H) were utilized after being cleaned using acetone, isopropyl alcohol, and distilled water. The substrates were held at a constant temperature of 100°C throughout the sputtering process, in which a power of 100 W and an argon atmosphere with 10 mtorr process pressure were used. Sputtering time of 90 min resulted in a 6-μm-thick Sb_2Te_3 and 5 μm Bi_2Te_3 films, respectively. Electrical contacts between the copper tape electrodes and the thermoelectric layer were made using adhesive conductive silver paste (42469, Alfa Aesar). The resistance between the electrodes for Sb_2Te_3 films and Bi_2Te_3 films ranged from 6 to 12 Ω and 2 to 10 Ω, respectively [8,20].

4.4.1.3　Characterization of Films

Figure 4.4(a) and 4.4(b) shows scanning electron micrographs of Sb_2Te_3 and Bi_2Te_3 films, respectively. The images clearly highlight the difference between the structures of the two films with Bi_2Te_3 exhibiting a more irregular structure compared to Sb_2Te_3. This confirms the polycrystalline nature of Bi_2Te_3. The SEM images also show that Bi_2Te_3 films are more porous than Sb_2Te_3 when they are deposited on Al_2O_3 substrates under identical conditions.

A surface profilometer (Ambios Technology XP-2) was used to investigate the morphology of the Sb_2Te_3 and Bi_2Te_3 surfaces and that of the substrate further. The average surface roughness of Sb_2Te_3 and Bi_2Te_3 was determined to be 0.4 and 0.7 μm, respectively. This indicates that the Bi_2Te_3 films are more porous, thereby confirming the earlier observation made from the micrograph in Figure 4.4(b). A greater porosity enables more fuel to be placed on the surface of the films. Additionally, a higher porosity is also important to facilitate increased thermal conduction between the fuel and thermoelectric films [2], resulting in a more sustained reaction propagation.

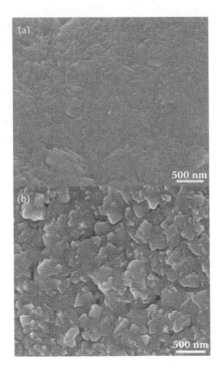

FIGURE 4.4　The SEM images of (a) Sb_2Te_3 film on Al_2O_3; (b) Bi_2Te_3 film on Al_2O_3.

X-ray diffraction measurements were carried out on the Sb_2Te_3 and Bi_2Te_3 samples in order to assess the crystallinity of the films. The XRD pattern for Sb_2Te_3 (Figure 4.5a [1]) shows three well defined peaks, which indicate its crystalline structure. In the case of Bi_2Te_3, the XRD pattern reveals a more polycrystalline structure (Figure 4.5b [1]) [21]. On addition of nitrocellulose, different effects on the XRD patterns of Sb_2Te_3 and Bi_2Te_3 are observed (Figure 4.5a and b [2]). For Bi_2Te_3, the peak at 18.1 2θ (006) plane for rhombohedral Bi_2Te_3 (ICDD no. 1-75-0921) is no longer exhibited in the samples with nitrocellulose (Figure 4.5b [1]). Possibly, incoherent scattering and adsorption effects may be responsible for the disappearance of this peak. It is also probable that the nitrocellulose coating on both the Sb_2Te_3 and Bi_2Te_3 samples effectively masks the surface, thus inhibiting the penetration of x-rays to and collection of scattering data from this particular peak. This effect is particularly acute for the relatively low-intensity (006) Bi_2Te_3 peak, which lies in the same region as the principal XRD reflections for nitrocellulose (ICDD no. 03-0114) [2]. Sb_2Te_3, on the other hand, does not show any significant diffraction peaks in this region.

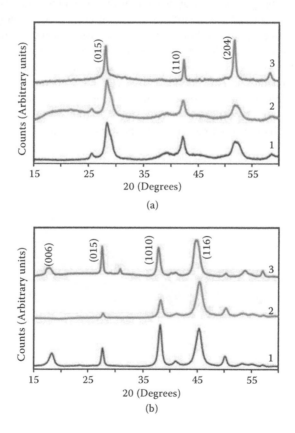

FIGURE 4.5   (a) The XRD pattern of Sb_2Te_3 (1) before thermopower wave propagation, (2) after adding nitrocellulose, and (3) after thermopower wave propagation. (b) The XRD pattern of Bi_2Te_3 (1) before thermopower wave propagation, (2) after adding nitrocellulose, and (3) after thermopower wave propagation.

Thus, the masking effect caused by the nitrocellulose layer only results in a general decrease in diffraction intensity across its entire diffraction pattern. The XRD patterns of the samples following consumption of the nitrocellulose are also shown (Figure 4.5a and b [3]) and will be discussed later [8].

4.4.2 RESULTS AND DISCUSSION

Figure 4.6 shows the schematic of the setup used for the thermopower experiments. Ignition takes place at one end of this system and the reaction self-propagates to the other end. A custom-made blow torch with a localized flame tip is used to initiate the reaction. Samples with low mass of nitrocellulose or NaN_3 (masses below 15 mg/75 mm^2) showed no sustained wave front propagation in any direction. Several samples using different masses of the fuels were prepared and tested.

For both Bi_2Te_3 and Sb_2Te_3 systems, identical masses were used to ensure statistical comparison. Ignition at one end of the system, shown in Figure 4.6, results in an accelerated self-propagating reaction wave that drives a simultaneous wave of free carriers. This wave of carriers results in a flow of current and voltage across the devices. The voltage signal exhibited a positive polarity for waves initiating at the positive electrode for p-type Sb_2Te_3 and the opposite polarity for the n-type Bi_2Te_3. The duration of the output signal corresponds to the reaction time, thus enabling us to calculate the reaction propagation velocity. Even in oscillatory mode, the thermopower waves always exhibit a constant polarity during the reaction, depending on the direction of wave front propagation. This suggests that the wave passes through the system faster than the cooling time of its posterior region [2,22].

FIGURE 4.6   Schematic of a thin film thermopower wave generation system.

4.4.3 Comparison of Sb₂Te₃- and Bi₂Te₃-Based Thermopower Devices

4.4.3.1 Thermopower Devices Based on Al₂O₃ Substrate

The thermal conductivity of Al_2O_3 is much higher than that of Bi_2Te_3 and Sb_2Te_3. Both Bi_2Te_3 and Sb_2Te_3 films are comparatively thin (<10 μm compared to the ~100 μm thick Al_2O_3 substrate). Thus, we can assume that the Al_2O_3 substrate will dominate the thermal conduction process [2].

Figure 4.7(a, b) shows typical voltage measurements across the fuel/Sb_2Te_3/Al_2O_3 and the fuel/Bi_2Te_3/Al_2O_3 systems, respectively. The voltage profiles of Figure 4.7(a, b) can be divided into an initial reaction phase followed by a cooling down phase. The reaction phase consists of a rising voltage and continues until all the combustion fuel is consumed.

This is followed by a region of exponential decay [15]. The voltage generated depends on the sample mass, especially the amount of fuel used (which is a combination of nitrocellulose and sodium azide). Sb_2Te_3-based devices generated voltages as large as 200 mV with oscillations in the range of 10–40 mV. On the other hand, Bi_2Te_3 devices generated voltages as large as 140 mV with larger oscillations generally within the range of 40–140 mV. Figure 4.7(a, b) also shows the difference in the oscillation amplitudes of the Sb_2Te_3 and Bi_2Te_3 devices. The reason for the difference in oscillatory behavior can be attributed to the larger thermal conductivity of Sb_2Te_3 (2.5 times that of Bi_2Te_3). Oscillation amplitude difference between the two devices corresponds to the difference in thermal conductivities of the two materials. A larger effective thermal conductivity means that the reaction will propagate faster, potentially affecting the oscillation amplitude.

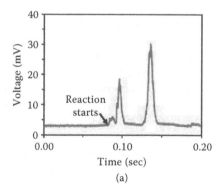 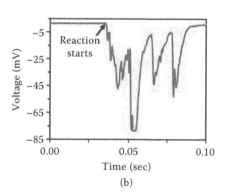

FIGURE 4.7 Comparison of oscillations between Sb_2Te_3- and Bi_2Te_3-based thermopower devices. (a) Oscillatory thermopower voltage signal for Sb_2Te_3 device using Al_2O_3 substrate. (b) Oscillatory thermopower voltage signal for Bi_2Te_3 device using Al_2O_3 substrate. We can see that Bi_2Te_3-based devices exhibit amplitudes approximately three times larger than those of Sb_2Te_3-based devices. The difference is similar to the variation in the thermal conductivities of Bi_2Te_3 and Sb_2Te_3.

The change in thermal conductivity affects the heat loss parameter as discussed in Section 4.2. The higher thermal conductivity of Sb_2Te_3 compared to Bi_2Te_3 enhances the heat transfer. Hence, faster combustion velocities are predicted for Sb_2Te_3-based devices and the system is closer to the parameter values where oscillations in the combustion occur. However, as the model is based on a one-dimensional average over all the layers, the resultant changes predicted are not as dramatic as observed in the experiments. Abrahamson et al. [15] have employed a similar averaged one-dimensional model. They have solved their model numerically (rather than use the asymptotic results that we have) with a set of estimated parameter values and found corresponding results to those reported in this work, albeit for carbon nanotube-guided thermopower waves, which are on a smaller scale than the thin-film-based devices.

The analysis of voltage and power as a function of mass shows an optimal mass range for both Sb_2Te_3- and Bi_2Te_3-based devices (Figure 4.8a, b). A very low amount of fuel does not generate enough heat to sustain the propagation of the exothermic reaction. Too much fuel, on the other hand, may provide more heat energy, but the unreacted accumulated fuel mass renders the reaction unsustainable [2]. As a result, we observed an optimal mass range (22–26 mg for ~75 mm^2 samples) that sustained the wave front propagation. Both types of devices exhibited well-matched trends in terms of the relation between voltage and mass (Figure 4.8a).

Comparison of peak power versus mass for the two different materials is shown in Figure 4.8(b). The peak power was determined using the equation

$$P = \frac{V^2}{R},\qquad(4.9)$$

in which P is the peak power in watts, V is the peak output voltage in volts, and R is the resistance in ohms.

The resistance of the Sb_2Te_3 films ranged from 6–12 Ω, while Bi_2Te_3 films' resistances were between 2 and 10 Ω. The difference between the resistances can be attributed to the larger electrical conductivity of Bi_2Te_3. Hence, we expect the power output from Bi_2Te_3-based devices to be higher. Experimental observations show that the peak power generated by Sb_2Te_3-based devices is ~6 mW, while Bi_2Te_3-based devices on an average generate 67% larger power with the peak value of ~10 mW. The resistance of the thermoelectric films remains almost the same even after the end of the reactions. The resistance dependence on temperature is defined by the following equation:

$$R(T) = R_o[1 + \alpha\,(T - T_o)]\qquad(4.10)$$

where T is the temperature, T_o is the reference temperature, R_o is the resistance at T_o, and α is the temperature coefficient of resistivity of the material. Since Sb_2Te_3 and Bi_2Te_3 are semiconductors, they exhibit negative temperature coefficient of resistivity [23,24]. Hence, it is expected that the resistance actually decreases during the reaction propagation in our case.

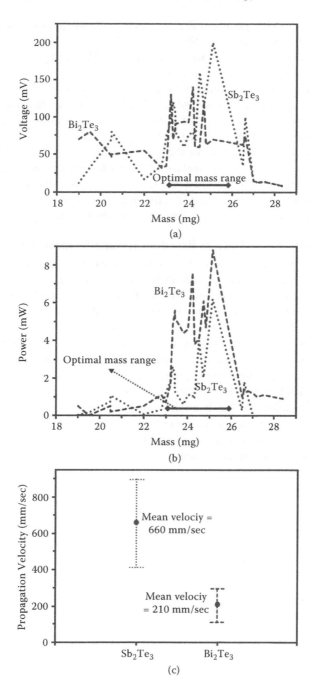

FIGURE 4.8 (a) Comparison of voltage generated as a function of mass (of fuel and thermoelectric material) for Sb_2Te_3- and Bi_2Te_3-based thermopower devices. (b) Power as a function of mass. (c) Comparison of the reaction propagation velocity for Sb_2Te_3- and Bi_2Te_3-based devices.

Additionally, the XRD pattern of the samples after the thermopower wave propagation, confirms that the structures of Sb_2Te_3 and Bi_2Te_3 remain unchanged (Figure 4.5a and b [3]), as the peaks remain dominantly unaltered. The average wave front propagation velocity for Sb_2Te_3-based devices was approximately 0.7 m/s compared to ~0.3 m/s average propagation velocity for Bi_2Te_3-based devices. Propagation velocities for Sb_2Te_3-based devices were generally 2–2.5 times higher than the velocity for Bi_2Te_3-based devices (Figure 4.8c). This nicely corresponds to the difference in the thermal conductivities of the two thermoelectric materials.

A higher thermal conductivity of the thermoelectric material provides an extra path for heat conduction (i.e., the surface of the Sb_2Te_3 film) in addition to the path provided by the thermally conductive Al_2O_3 substrate. This will consequently cause the wave to travel faster on Sb_2Te_3, as the thermal wave generated from the exothermic reaction travels approximately 2.5 times faster than on Bi_2Te_3.

4.4.3.2 Thermopower Devices Based on Terracotta Substrate

The oscillatory behavior and wave front velocities strongly depend on the thermal conductivity of the substrate. Terracotta has an identical thermal conductivity to Bi_2Te_3. However, thermal conductivity of Sb_2Te_3 is 2.5 times higher. Consequently, we expect the rate of heat conduction for Sb_2Te_3-based devices to be approximately 2.5 times faster than for Bi_2Te_3-based thermopower devices.

Figure 4.9(a,b) shows the voltage signals obtained for Sb_2Te_3/terracotta and Bi_2Te_3/terracotta systems, respectively. In line with our expectation, we can see that the reaction propagation for the Sb_2Te_3/terracotta device is approximately three times faster than the Bi_2Te_3/terracotta device. This endorses our explanation that the reaction propagation velocity depends on the thermal conductivity of the material. The voltages generated by both devices are comparatively low because a relatively high thermal conductivity is essential for a sustained thermopower wave propagation and to maintain a high reaction temperature [2].

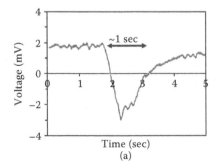 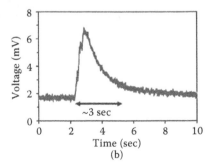

FIGURE 4.9 Comparison of thermopower voltage signals on terracotta substrates for (a) Sb_2Te_3 films; (b) Bi_2Te_3 films.

4.5 ZnO-BASED THERMOPOWER WAVE SOURCES

4.5.1 FABRICATION OF ZnO THIN FILMS

The ZnO films were deposited using RF magnetron sputtering. A high-purity (99%) ZnO sputtering target (Vin Karola Instruments) was used in the process. The sputtering process was started after the chamber was pumped down to 2×10^{-5} Torr. Al_2O_3 substrates were used after being adequately cleaned using acetone, isopropanol, and distilled water successively and dried using a high-pressure nitrogen gun. The substrates were kept at room temperature during the sputtering process. RF sputtering power of 160 W was used throughout the deposition process. The sputtering process was carried out under an atmosphere of 100% argon (Ar).

A sputtering time of 60 min resulted in ~1 µm ZnO layer. Adhesive conductive silver (Ag) paste (42469, Alfa Aesar) was used to make electrical contacts between the copper tape electrodes and the ZnO samples. ZnO films exhibited a room temperature resistance of approximately 3 kΩ [9]. However, as ZnO exhibits a relatively high negative temperature coefficient of resistance, the resistance is considerably lower at the reaction temperature. This property will be discussed later in the chapter.

4.6 RESULTS AND DISCUSSION

Zinc oxide, an N-type semiconducting oxide, is an outstanding candidate as the core thermoelectric material for thermopower-based energy sources. ZnO exhibits a relatively high Seebeck coefficient (approximately −360 µV/K at 85°C), which increases with temperature, and a high electrical conductivity at elevated temperatures of above 300°C (ZnO film resistance of 250 Ω at 300°C—hence, it is suitable for operation in thermopower wave sources), high thermal conductivity (15 W/m.K at 300°C), and good chemical stability [25,26]. The temperature dependence of the Seebeck coefficient and electrical conductivity is discussed in this chapter.

It has been shown that ZnO-based thermopower devices are capable of generating significantly larger voltages and oscillation amplitudes than previously reported systems utilizing MWNTs, Bi_2Te_3, and Sb_2Te_3 as the core thermoelectric materials.

The ZnO films were deposited on alumina (Al_2O_3) substrates using RF magnetron sputtering. Previous work carried out on RF sputtered ZnO films shows that their resistivity can vary up to eight orders of magnitude depending on the sputtering power and the substrate temperature [27-29]. Consequently, a high sputtering power (160 W) was used and no substrate heating was applied in order to synthesize crystalline and electrically conductive ZnO thin films (~1.2 µm). Nitrocellulose [$C_6H_8(NO_2)_2O_5$] was used as the fuel, due to its large enthalpy of reaction (4.75×10^6 J/kg) [30]. Sodium azide (NaN_3) in aqueous solution (50 mg/mL) was then added to serve as a primary igniter to lower activation energy (40 kJ/mol for NaN_3 compared to 110–150 kJ/mol for nitrocellulose).

Similarly to Bi_2Te_3- and Sb_2Te_3-based thermopower wave systems, ignition is initiated at one end using a blow torch with a fine tip, resulting in self-propagating reaction waves that travel at a rapid pace to the opposite end. This accelerated reaction

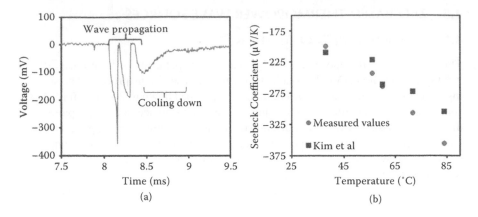

FIGURE 4.10 (a) Oscillatory thermopower voltage signal obtained using the fuel/ZnO/Al₂O₃ device. (b) Measured and literature values of the Seebeck coefficient at different temperatures [25].

wave entrains a simultaneous wave of electrical carriers, resulting in an oscillatory voltage output.

A typical thermopower voltage signal obtained across a sample of the fuel/ZnO/Al₂O₃ system is shown in Figure 4.10(a). The moving temperature gradient results in voltages with peak magnitudes of up to 500 mV and oscillations with peak-to-peak amplitude of up to 400 mV. The power obtained from these devices can be as large as 1 mW.

Using the theoretical model presented in Section 4.2, we were able to predict the oscillatory behavior and the propagation velocity of the thermopower waves [10]. The theoretical analysis concluded that the oscillatory behavior and wave front velocities strongly depend on the thermal conductivity. A change in thermal conductivity affects the heat loss parameters of the system. It was demonstrated that a higher thermal conductivity enhances the heat transfer through the thermoelectric core, resulting in a proportional increase in combustion velocities. Consequently, Sb_2Te_3-based systems exhibited combustion velocities that were almost 2.5 times faster than the Bi_2Te_3-based devices, reflecting the difference in the thermal conductivities of the two materials [8].

Hence, based on the aforementioned analysis, it is expected that a higher thermal conductivity of ZnO (~15.0 W/m.K) compared to Bi_2Te_3 (1.0 W/m.K) and Sb_2Te_3 (2.5 W/m.K) would result in a faster wave velocity and larger oscillations. The velocity of the waves in our system was assessed using the duration in which the output voltage exists. We have previously shown that the duration of the voltage signal corresponds to the reaction wave propagation time, which enables us to calculate the linear propagation velocity of the thermopower waves [8,20]. A high thermal conductivity of the ZnO thermoelectric material provides an extra path for heat conduction (i.e., the surface of the ZnO film) in addition to the path provided by the thermally conductive Al₂O₃ substrate. As a result, we initially expect the rate of heat conduction for ZnO-based devices to be approximately 15 times faster than for Bi_2Te_3-based thermopower devices. Propagation velocities for the ZnO-based devices were measured to range between 12 and 35 m/s, which was generally 15–25 times higher than the velocities for Sb_2Te_3- (1.2 m/s) and Bi_2Te_3- (0.4 m/s) based devices demonstrated previously [8,20]. This difference nicely corresponds to the variation in the thermal conductivities of these materials and is in excellent agreement with theoretical expectations.

It has been previously shown that in a thin film thermopower system configuration in which nitrocellulose is incorporated as the fuel, the reaction temperature remains under 300°C [8,20]. As a result, the Seebeck coefficient and resistance of the ZnO films within this range needed to be estimated. Kim et al. have shown that the Seebeck coefficient of ZnO increases with temperature. In their work, it has been demonstrated that the Seebeck coefficient of ZnO is approximately −300 μV K^{-1} at 85°C [25]. The Seebeck coefficient of the ZnO films was measured to be approximately −360 μV K^{-1} (Figure 4.10b). At 300°C, the Seebeck coefficient is expected to be approximately −500 μV K^{-1} based on linear extrapolation [25]. It has been reported that ZnO exhibits a negative temperature coefficient of resistance at temperatures up to 350°C [31,32]. The ZnO film resistance was measured at temperatures of up to 300°C, and the value of the temperature coefficient of resistance is in close agreement with Caillaud, Smith, and Baumard (approximately −3.5 × 10^{-3}/°C) [31]. At 300°C, the resistance was reduced to approximately 250 Ω (from 3.4 kΩ at room temperature). Figure 4.11(a) illustrates the Raman spectra (indicated as "1") of the as-grown ZnO structure. The peaks observed at around 436 and 576/cm correspond to the E_2 (high) and the dominant A_1 (LO) modes, respectively. The E_2 (high) is a characteristic mode of the ZnO hexagonal wurtzite-type lattice [33]. The Raman peaks after the deposition of nitrocellulose (spectrum "2" in Figure 4.11a) show the presence of strong C–H and NO_2 bonds [34].

After the thermopower wave propagation, the C–H and NO_2 bonds disappear, while carbon peaks appear on the Raman spectra (spectrum 3 in Figure 4.11a). The exothermic reaction causes the C–H bonds to decompose, while carbon is left behind as the residual product. The x-ray diffraction pattern in Figure 4.11(b) reveals that the ZnO films are highly crystalline with a strong c-directional (002) preferred orientation. Addition of nitrocellulose results in an acute reduction in intensity of the 25.3° 2θ Al_2O_3 substrate peak. This is because the 25.3° 2θ Al_2O_3 substrate peak lies in the same region as the principal XRD reflections for nitrocellulose (ICDD no. 03-0114). A general decrease in diffraction intensity due to the nitrocellulose coating inhibiting x-ray penetration is observed [8]. The XRD pattern following thermopower wave propagation shows strong ZnO peaks (Figure 4.11b), verifying that ZnO remains largely intact at the end of the process.

Using AFM, the roughness of the ZnO films was determined to be approximately 250 nm (Figure 4.12).

It has therefore been shown that ZnO is excellent for application as a core material for thermopower wave-based energy sources as it exhibits a high Seebeck coefficient (−360 μV/K), electrical conductivity (250 Ω resistance at 300°C), and thermal conductivity (15 W/m.K) [25]. The aforementioned properties of ZnO resulted in a highly oscillatory voltage output of the order of 500 mV and peak specific power of the order of 0.5 kW/kg.

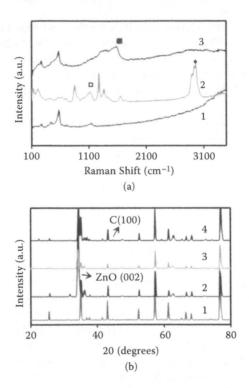

FIGURE 4.11 (a) Raman spectra of (1) ZnO film, (2) nitrocellulose-coated ZnO film (■ peak denotes strong NO_2 bonds; ◆ denotes strong C–H bonds), and (3) after the reaction (□ peak denotes carbon). (b) XRD pattern (1) of Al_2O_3 substrate, (2) of ZnO before thermopower wave propagation, (3) after adding nitrocellulose, and (4) after thermopower wave propagation.

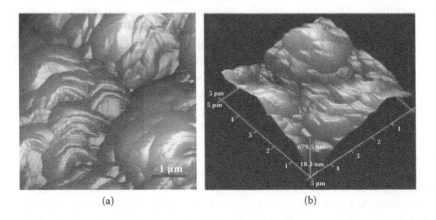

FIGURE 4.12 Samples of AFM images: (a) two-dimensional image of the ZnO film surface; (b) three-dimensional profile of the ZnO film surface

4.7 CONCLUSION

It has been shown that thermopower waves offer a truly unique method of scaling down the dimensions of energy sources while maintaining high specific powers. Thermopower wave systems appear to be an attractive technology for micro- and nanoscale energy generation systems, as their efficiency and the energy discharge rates increase with a reduction in dimensions.

Such systems therefore provide an excellent opportunity for future small size sources that are compatible with low-dimensional electronics.

Furthermore, such systems currently use solid fuels for the exothermic chemical reaction. Replacing solid fuels with liquid fuels can help integrate such sources with micro- and nanofluidic platforms, making a novel combination. This will transform the thermopower wave energy generation systems into a replenishable and sustainable energy generation technology.

REFERENCES

1. Cook, B. W., Lanzisera, S., and Pister, K. S. J. 2006. SoC issues for RF smart dust. *Proceedings of IEEE* 94:1177–1196.
2. Walia, S., Weber, R., Latham, K., Abrahamson, J. T., Strano, M. S., and Kalantar-zadeh, K. 2011. Oscillatory thermopower waves based on Bi$_2$Te$_3$ films. *Advanced Functional Materials* 21:2072–2079.
3. Sekido, S. 1983. Solid state micro power sources. *Solid State Ionics* 9 (10): 777–781.
4. Murphy, O. J., Cisar, A., and Clarke, E. 1998. Low-cost light weight high power density PEM fuel cell stack. *Electrochimica Acta* 43:3829–3840.
5. Martin, J. L., and Osenar, P. 2009. Portable military fuel cell power systems. *ECS Transactions* 25: 249–257.
6. Nakanishi, H., and Grzybowski, B. A. 2010. Supercapacitors based on metal electrodes prepared from nanoparticle mixtures at room temperature. *Journal of Physical Chemistry Letters* 1:1428–1431.
7. Choi, W., Abrahamson, J. T., Strano, J. M., and Strano, M. S. 2010. Carbon nanotube-guided thermopower waves. *Materials Today* 13:22–33.
8. Walia, S., Weber, R., Sriram, S., Bhaskaran, M., Latham, K., Zhuiykov, S., et al. 2011. Sb2Te3 and Bi2Te3 based thermopower wave sources. *Energy & Environmental Science* 4:3558–3564.
9. Walia, S., Weber, R., Balendhran, S., Yao, D., Abrahamson, J. T., Zhuiykov, S., et al. 2012. ZnO based thermopower wave sources. *Chemical Communications* 48:7462–7464.
10. Mercer, G. N., Weber, R. O., and Sidhu, H. S. 1998. An oscillatory route to extinction for solid fuel combustion waves due to heat losses. *Proceedings of the Royal Society of London, Series A—Mathematical Physical and Engineering Sciences* 454:2015–2022.
11. Zarko, V., and Kiskin, A. 1980. Numerical modeling of nonsteady powder combustion under the action of a light flux. *Combustion, Explosion, and Shock Waves* 16:650–654.
12. Bayliss, A., and Matkowsky, B. J. 1994. From traveling waves to chaos in combustion. *SIAM Journal of Applied Mathematics* 54:147–174.
13. McIntosh, A. C., Weber, R. O., and Mercer, G. N. 2004. Nonadiabatic combustion waves for general Lewis numbers: Wave speed and extinction conditions. *Anziam Journal* 46:1–416.
14. Choi, W., Hong, S., Abrahamson, J. T., Han, J. H., Song, C., Nair, N., et al. 2010. Chemically driven carbon–nanotube-guided thermopower waves. *Nature Materials* 9:423–429.

15. Abrahamson, J. T., Choi, W., Schonenbach, N. S., Park, J., Han, J.-H., Walsh, M. P., et al. 2010. Wavefront velocity oscillations of carbon-nanotube-guided thermopower waves: Nanoscale alternating current sources. *ACS Nano* 5:367–375.
16. Sommer, I. 1972. Crystal growth of Sb_2Te_3 by chemical transport. *Journal of Crystal Growth* 12:259–260.
17. Zimmer, A., Stein, N., Johann, L., Terryn, H., and Boulanger, C. 2008. Characterizations of bismuth telluride films from Mott-Schottky plot and spectroscopic ellipsometry. *Surface and Interface Analysis* 40:593–596.
18. Burnham, A. K., and Fried, L. E. 2006. Kinetics of PBX9404 aging (http://www.ntis.gov/search/product.aspx?ABBR=DE2006894349).
19. Katoh, K., Ito, S., Kawaguchi, S., Higashi, E., Nakano, K., Ogata, Y., et al. 2010. Effect of heating rate on the thermal behavior of nitrocellulose. *Journal of Thermal Analysis and Calorimetry* 100:303–308.
20. Walia, S., Weber, R., Latham, K., Petersen, P., Abrahamson, J. T., Strano, M. S., et al. 2011. Oscillatory thermopower waves based on Bi2Te3 films. *Advanced Functional Materials* 21:2072–2079.
21. Pradyumnan, P. P., Swathikrishnan. 2010. Thermoelectric properties of Bi2Te3 and Sb2Te3 and its bilayer thin films. *Indian Journal of Pure & Applied Physics* 48:115–120.
22. Choi, W., Hong, S., Abrahamson, J. T., Han, J. H., Song, C., Nair, N., et al. 2010. Chemically driven carbon-nanotube-guided thermopower waves. *Nature Materials* 9:423–429.
23. Das, V. D., Soundararajan, N., and Pattabi, M. 1987. Electrical conductivity and thermoelectric power of amorphous Sb_2Te_3 thin films and amorphous crystalline transition. *Journal of Materials Science* 22:3522–3528.
24. Das, V. D., and Soundararajan, N. 1988. Size and temperature effects on the thermoelectric power and electrical resistivity of bismuth telluride thin films. *Physical Review B* 37:4552–4559.
25. Kim, K. H., Shim, S. H., Shim, K. B., Niihara, K., and Hojo, J. 2005. Microstructural and thermoelectric characteristics of zinc oxide-based thermoelectric materials fabricated using a spark plasma sintering process. *Journal of the American Ceramic Society* 88:628–632.
26. Jood, P., Mehta, R. J., Zhang, Y. L., Peleckis, G., Wang, X. L., Siegel, R. W., et al. 2011. Al-doped zinc oxide nanocomposites with enhanced thermoelectric properties. *Nano Letters* 11:4337–4342.
27. Lim, S. J., Kwon, S. J., and Kim, H. 2008. ZnO thin films prepared by atomic layer deposition and rf sputtering as an active layer for thin film transistor. *Thin Solid Films* 516:1523–1528.
28. Jayaraj, M. K., Antony, A., and Ramachandran, M. 2002. Transparent conducting zinc oxide thin film prepared by off-axis RF magnetron sputtering. *Bulletin of Materials Science* 25:227–230.
29. Martínez, M. A., Herrero, J., and Gutiérrez, M. T. 1994. Properties of RF sputtered zinc oxide based thin films made from different targets. *Solar Energy Materials and Solar Cells* 31:489–498.
30. Lemieux, E., and Prud'Homme, R. E. 1985. Heats of decomposition, combustion and explosion of nitrocelluloses derived from wood and cotton. *Thermochimica Acta* 89:11–26.
31. Caillaud, F., Smith, A., and Baumard, J.-F. 1991. Effect of oxygen chemisorption on the electrical conductivity of zinc oxide films prepared by a spray pyrolysis method. *Journal of the European Ceramic Society* 7:379–383.
32. Al-Hardan, N. H., Abdullah, M. J., and Aziz, A. A. 2010. Sensing mechanism of hydrogen gas sensor based on RF-sputtered ZnO thin films. *International Journal of Hydrogen Energy* 35:4428–4434.

33. Ozgur, U., Alivov, Y. I., Liu, C., Teke, A., Reshchikov, M. A., Dogan, S., et al. 2005. A comprehensive review of ZnO materials and devices. *Journal of Applied Physics* 98:041301.
34. Moore, D. S., and McGrane, S. D. 2003. Comparative infrared and Raman spectroscopy of energetic polymers. *Journal of Molecular Structure* 661:561–566.

5 Polymer Solar Cell
An Energy Source for Low-Power Consumption Electronics

*Badr Omrane, Sasan V. Grayli, Vivien Lo,
Clint Landrock, Siamack V. Grayli, Jeydmer
Aristizabal, Yindar Chuo, and Bozena Kaminska*

CONTENTS

Abstract: The thin, flexible, and low-cost nature of polymer solar cells (PSCs) makes them a suitable energy source for low-power consumption electronics. Recent technological advances have considerably enhanced the power conversion efficiencies and lifetimes of PSCs. This chapter presents an overview of recent technological advances in PSCs, and their integration into large-scale fabrication processes as well as their impact on the cost of the final products.

5.1 INTRODUCTION

Our ever increasing global demand for energy and subsequent resource consumption has emphasized a need for more efficient and sustainable power generation technologies. However, the largest roadblock to widespread adoption of any type of environmentally sustainable energy is cost. As a result, continuous efforts in product and process innovation are underway to increase affordability and availability of "green" energy. Photovoltaics are perhaps the most attractive energy producers as the sun is the most accessible and sustainable energy source found on Earth. Silicon (Si)-based photovoltaics (PVs) can achieve high-power conversion efficiency (PCE) with an average life span of 20 to 30 years, but their high manufacturing costs and rigid, fragile structure discourage greater use. Currently, Si-PVs account for 13% of market share in the global renewable energy sector and a mere 3% of the total global energy generation sector [1]. In the last decade, several viable lower cost alternatives to Si-PVs have been developed. These include a number of thin-film technologies such as copper–indium–gallium–selenium (CIGS) PVs, dye-sensitized PVs, amorphous silicon PVs, and organic polymer PVs.

One of the most important challenges in sensors and systems deployed in wireless, portable, or wearable applications is designing and selecting an appropriate energy system. For most electronics, the energy required to power the microprocessor is negligible; however, the sensor and actuator elements may consume significant amounts of power. In applications that require ultralightweight systems, such as medical applications for remote care and home monitoring, often the primary energy source is a battery that is substantially larger than the system it powers in order to meet the requirements for wireless use.

Recent developments in biomedical devices demonstrate power consumption measured in the milliwatt and even nanowatt range. Marzencki et al. [2] have presented a wearable wireless sensor system requiring 2.1 V operating voltage with 1.86 to 16.6 mW power used in conjunction with a mobile phone for data collection; Xiaodan et al. [3] have reported on a 1 V sensor interface chip requiring a mere 450 nW of power. In order to take advantage of these devices' small footprint and flexible nature, a powering system that is comparably small and flexible is needed. For wearable systems, in particular, the battery footprint becomes a very important issue. Recent efforts to reduce this battery footprint along with increasing device lifetimes and reliability have included supplementary power sources using solar or mechanical energy harvesting. Integrated solar cells on outdoor wireless sensor nodes and vibration energy harvesters used on automotive sensor units are some commonly found examples. Assisted powering of autonomous electronic devices with energy harvesting is a nontrivial challenge that requires the matching of function and environment, and, hence, there is no universal solution (e.g., photovoltaic energy harvesting is not suitable for a subcutaneous implantable biosensor).

Of solar harvesting technologies available today, polymer solar cells (PSCs) offer the most flexibility in manufacturing processes and, thus, ease of integration. With thicknesses of the PV layers only in tens of nanometers, PSCs require very little material and can potentially be recycled. PSC inks can be printed on thin, flexible substrates, such as plastics, paper, metal siding, roofing tiles, etc., using roll-to-roll,

TABLE 5.1

Cost Breakdown of Primary Materials for the Proposed Energy Harvesting System Compared with Current Energy Harvesting and Storage Devices

Materials	Raw Material ($/m²)	Processing/Overhead ($/m²)	Total ($/m²)	Total Cost ($/watt)
PSC[a]	23.40	23.08	46.48	<1.0
mc-Si[a]	208.00	135.50	343.50	2.34
Li ion battery[b]			N/A	0.77

[a] Kalowekamo, J. and Baker, E. (2009). *Solar Energy* 83:1224–1231.
[b] Wee, J. (2007). *Journal of Power Sources* 173:424–436.

ink jet, spray coating, and many other scalable print-based manufacturing methods. Recent advances in polymer solar cell technology have increased the PCE to higher than 12% (Heliatek GmbH), and efficiencies could reach as high as 17% in the near future [4]. Combined with the low costs of manufacturing, PSCs can potentially offer a very attractive subdollar-per-watt figure and are anticipated to greatly surpass silicon-based solar cells in both use and application range.

Despite their exciting outlook, a number of challenges hinder the commercial use of PSCs. Stability of PSC organic molecules has been elusive and greatly limits the lifetime and performance of PSCs. Oxygen, in particular, rapidly degrades the photovoltaic function, requiring manufacturing to be run in an inert environment and PSCs to be packaged in protective encapsulates. The primary focus of PSC research today is to address challenges in overcoming high manufacturing costs due to the need for specialized equipment, as well as performance issues that are limiting wide use. An additional challenge in realizing the full potential of PSCs is that they must be integrated with an efficient energy storage system that is as thin, lightweight, and flexible as the PSC. Table 5.1 gives the cost estimates for the energy harvesting system components in comparison to monocrystalline solar cells [5] and Li-ion batteries [6]. The costs for PSC components assume roll-to-roll manufacturing and processing in an ambient environment. The cost per unit area of producing the devices is well below a dollar per square centimeter, and the cost per power generated is well below that of state-of-the-art technologies available today.

This chapter reviews recent technological advances in PSCs, including novel polymer materials and nano-optics. The integration of PSCs into large-scale fabrication processes as well as their impact on the cost of the final products is also reported.

5.2 POLYMER SOLAR CELLS: THEORY CONSIDERATIONS AND SURVEY ON EXISTING AND NEW POLYMERS

Classically, PSCs are made of a thin film of a semiconductor material inserted between two electrodes, with one of the electrodes being transparent. The semiconductor material is typically either a bulk-heterojunction (BHJ) blend or a bilayer of

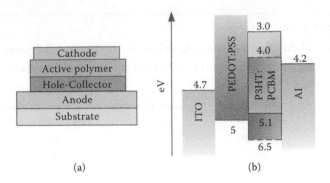

FIGURE 5.1 Schematic of (a) a standard PSC configuration, and (b) energy diagram under flat band condition of a conventional P3HT:PCBM PSC.

an active electron donor polymer and an electron acceptor polymer (Figure 5.1a). Under illumination, excitons are created in the donor region. Due to the electric field built by the mismatch of the Fermi levels of the two electrodes (Figure 5.1b), the excitons are able to diffuse to the junction and separate into holes and electrons. The holes remain in the donor, while the electrons are passed to the acceptor. It is also common to have an interlayer made of a conductive material between the semi-conductor layer and the anode in order to improve charge transfer to the electrodes and, in some cases, to smooth the indium tin oxide (ITO) surface. Depending on the configuration of the PSC, this interlayer is usually referred to as a *hole collector* or an *electron blocking layer.*

To enhance the power conversion efficiency of a PSC, the quality and/or the nature of the electrodes should be improved, and the ratio of the number of charge carriers collected by the PSC to the number of incident photons should also be increased. This ratio is known as external quantum efficiency (EQE) and gives the amount of current density (J_{sc}) that a PSC can generate under a certain light spectrum (Φ):

$$J_{sc} = \frac{q}{hc} \int \lambda \times EQE(\lambda)\Phi(\lambda)d\lambda \qquad (5.1)$$

where λ, q, h, and c are the wavelength, electron charge, Plank's constant, and speed of light in vacuum, respectively.

The reliance on ITO as a transparent electrode remains a key factor in the performance of PSCs. More precisely, the trade-offs between conductivity and transmission, which depend highly on the method of deposition [7], as well as cracking and release of destructive oxygen atoms into the active layer, are major problems inherent to ITO [8]. Although transparent films of carbon nanotubes [9,10] or highly conductive polymers [11] have been proposed as replacements to ITO, the PCE of PSCs has not been drastically enhanced.

Ideally, conjugated polymers should exhibit low bandgaps to broaden the absorption range and a crystalline structure to enhance charge mobility. Other factors, such as the separation between the highest occupied molecular orbital (HOMO) of the donor and the lowest unoccupied molecular orbital (LUMO) of the acceptor, play a role in choice of the polymeric material used for PSC. The HOMO and LUMO of the PSC semiconductor materials can be tuned by modifying their backbones or side groups. The focus for low bandgap materials is mainly on those organic materials with an absorption range extended into the infrared. It is evident that there is a substantial gain in current density if the bandgap is decreased via an increase of the cut-off wavelength from 650 to 1000 nm [12,13]. In theory, lowering the bandgap of a polymer blend means increasing the strength of both the donor and the acceptor. In practice, the latter can be achieved through use of electron-withdrawing groups on the acceptor and electron-donating groups on the donor [12]. Some examples of electron-withdrawing groups are chemicals such as CN, NO^2, quinoxalines, and pyrozines; examples of electron-donating groups are thiophene and pyrrole [12].

The use of bulk heterojunction configuration can further enhance the maximum theoretical current density of the PSC. For a PSC to reach maximum power conversion efficiency the open-circuit voltage (V_{OC}) needs also to be considered [14]. The latter can be achieved, for instance, by increasing the acceptor LUMO in combination with low bandgap materials. This process is referred to as *energy level alignment*, and the range of energy levels for the acceptor is usually fixed by the fullerene materials. For poly(2-methoxy-5-ethylhexyloxy-1,4-phenylenevinylene) (MEH-PPV) and PCBM, the alignment between the LUMO can result in an efficient charge transfer. The charge transfer can become more efficient if MEH-PPV is replaced with poly(3-hexylthiophene) (P3HT) since the difference in energy levels is higher for P3HT compared to MEH-PPV, at the expense of a lower (V_{OC}) [12]. Over the past decades, a wide variety of electron donor materials that includes polymers and small molecular compounds has been developed. P3HT is one of the most common and widely used polythiophenes. The advantage of P3HT lies within the intrinsic directionality property of the thiophene ring that allows regularity in the structure where all the molecules can add in a head-to-tail fashion [12]. P3HT is a regioregular polymer, with a maximum absorption wavelength varying with the percentage of head-to-tail couplings, and is also dependent on the molecular weight.

The power conversion efficiency depends highly on how the polymer materials are being deposited. The first generations of PSCs were based on a bilayer structure in which the donor and acceptor materials were deposited on top of each other. In this configuration, the charge separation occurs at the interface between the donor and acceptor, which helps minimize carrier recombination. However, in a bilayer structure, the charge transfer between two layers is limited by the low surface area of the contact between the donors and acceptors. The limited donor–acceptor contact can be solved by using bulk heterojunction materials where the donor and the acceptor material are mixed together, creating a contact between the donor and the acceptor throughout the bulk of the material. The contact surface area between the donor and the acceptor is several orders of magnitude larger than that of the bilayer systems. Currently, the best choices of materials for BHJ PSCs are a combination of fullerene compounds with p-type polyconjugated polymers. For BHJ P3HT:PCBM cells, PCEs near 5% have been achieved compared to only 1%–2% PCE for the bilayer configurations [15,16].

TABLE 5.2

Device Characteristics for New Polymer Materials

Materials	J_{sc} (mA/cm²)	V_{oc} (V)	FF	PCE	Ref.
PBDTTT-CF[a]	15.2	0.76	0.669	7.73	50
PCPDTBT[b]	16.2	0.62	0.55	5.5	51
PCDTBT[c]	10.6	0.88	0.66	6.1	52
PBDTDTBT[d]	10.7	0.92	0.57	5.66	53
P3HT/IC60BA[e]	10.61	0.84	0.73	6.48	54

[a] Poly[4,8-bis-substituted-benzo[1,2-b:4,5-b0]dithiophene–2,6-diyl-alt–4-substituted-thieno[3,4-b]thiophene–2,6-diyl].
[b] Poly[2,1,3-benzothiadiazole–4,7-diyl[4,4-bis(2-ethylhexyl)–4H-cyclopenta[2,1-b:3,4-b′]dithiophene–2,6-diyl]].
[c] Poly[[9-(1-octylnonyl)–9H-carbazole–2,7-diyl]–2,5-thiophenediyl–2,1,3-benzothiadiazole–4,7-diyl–2,5-thiophenediyl].
[d] Poly[4,8-bis(2-ethylhexyl–2-thenyl)-benzo[1,2-b:4,5-b′]dithiophene-alt–5,5′-(4′,7′-di–2-thienyl–2′,1′,3′-benzothiadiazole)].
[e] Indene-C60 bis-adduct.

Even though over the past decade P3HT has been established as the most effective electron donor for PSCs, new materials with longer absorption range have been synthesized. Some examples of new polymers meeting the aforementioned criterion are reported in Table 5.2.

P3HT:PCBM-based PSCs have been susceptible to rapid degradation when exposed to an ambient environment. New materials such as PCDTBT have improved PSC stability in air; however, the fabrication process still requires an inert environment to avoid oxygen and moisture trapping within the layers. Recently, P3HT:PCBM-based PSCs, using an ITO anode and an indium metal cathode fabricated, stored, and tested entirely in air without encapsulation, exhibited less than 10% loss in power conversion efficiency after 200 days [17]. The fabrication process of indium-based PSCs is depicted in Figure 5.2(a) and does not require any physical vapor deposition

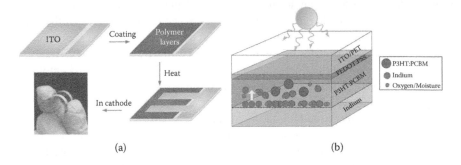

(a)   (b)

FIGURE 5.2 (a) Schematic of the fabrication process of indium-based PSCs, and (b) a simplified diagram proposed model of indium interaction with oxygen atoms and moisture. (Omrane, B. et al. 2011. *Applied Physics Letters* 4:263305.)

system since the indium cathode is deposited using a heat-press system in air. X-ray photoelectron spectroscopy shows that this improvement in ambient stability is correlated with the diffusion of indium from the cathode into the active polymer (Figure 5.2b). Diffusion and formation of In_2O_3 within the P3HT:PCBM decreases the amount of oxygen within the device available to oxidize the active polymer layer [17].

5.3 POLYMER SOLAR CELLS: NANO-OPTICS FOR ENHANCING EFFICIENCY

5.3.1 GRATING STRUCTURES

Subwavelength geometries and the generation of surface plasmon polaritons (SPPs) is a rapidly developing area of research in photonics [18], recently targeting potential applications for solar cells [19–23]. Maximizing the absorption of the incident light in PSCs is the key to improving their efficiency. To optimize PSC cells, two considerations need to be taken into account:

* Reducing the charge carrier transport paths by decreasing the active layer thickness in order to reduce carrier scattering
* Increasing active layer thickness to promote higher photon absorption [24–26]

Light trapping is one of the techniques used for increasing the absorption of photons in PSCs. The structures or layers that cause the light to be reflected internally can result in light trapping, which increases photon absorption and consequently improves the power conversion efficiency of PSCs [24,27,28]. One of the reported configurations for light trapping consists of a transparent polymer microprism with V-profiled grating structures on the surface. The PEDOT:PSS and the photoactive layer are sandwiched between the microprism substrate and the top electrode [24,28]. Figure 5.3 demonstrates the aforementioned structure.

Light trapping can also be achieved by using diffraction gratings within the PSC cells. These structures can enhance and improve the power conversion efficiency of the PSCs by increasing the absorption of the incident light [27]. In such configurations, the optical system of the PSC cell consists of a waveguide (i.e., diffraction grating) and a light guide (i.e., glass substrate). The refractive index of the substrate material defines the paths for the propagating diffracted orders of the incident light [27]. Niggemann et al. [27] report that the glass light-guiding effects are noticeable

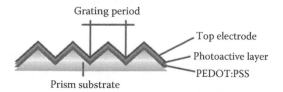

FIGURE 5.3 Schematic of a prism structure used to induce light trapping. (Adapted from Niggemann, M. et al. 2008. *Physica Status Solidi* 205:2862–2874.)

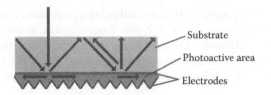

FIGURE 5.4   Schematic of a PSC with light trapping structures. (Adapted from Niggemann, M. et al. 2004. *Thin Solid Films* 451-452:619–623.)

for wavelengths larger than 650 nm where the transverse electrical (TE) polarization is absorbed by the active layer. Figure 5.4 depicts a PSC with a grating structure where the substrate acts as a light guide.

The use of reflective structures in PSCs is widely suggested due to the importance of light trapping effects for improving solar cells. Diffraction gratings not only can diffract light, but also can reflect diffracted light back into the active area. The different refractive index of the polymer used in a PSC cell and air can also cause reflection of the outgoing rays of light at the air–polymer interface, which in turn increases the chance of photon absorbance by the active layer of the cell [26,27]. In addition, Dal Zilio et al. [26] showed that by implementing a microlens system to a PSC cell, the efficiency can be improved by up to 25%. However, the presence of a reflective layer should not be neglected [26]. Diffraction gratings can also increase the absorption for the wavelengths around 500 nm as well as the near infrared (NIR) region for PSC using low-bandgap polymers. These types of polymers are known to have poor light absorption around the aforementioned wavelengths [29].

5.3.2   Surface Plasmon and Plasmonic Structures

With recent advancements in nanotechnology-related fields, harvesting surface plasmons (SPs) [18] and fabrication of structures that induce surface plasmon resonance (SPR) have become within reach for an easy integration with PSC [30]. This phenomenon was theorized and later introduced as a way to improve the power conversion efficiency of PSCs. This improvement is attributed to the enhanced optical field caused by SP structures that significantly increase the light absorption in the active polymer layer [31]. Diffraction gratings covered with a thin metallic layer not only act as a waveguide but can also generate SPR from the incident light. The generated SPR results in enhancement of the photoelectric conversion in the active layer, which translates into an increase of the PSC short-circuit current [31]. In Figure 5.5, a basic incorporation of grating coupled plasmonic structures in PSCs is illustrated.

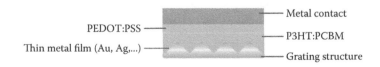

FIGURE 5.5   Schematic of a grating coupled surface plasmonic PSC.

There are other methods in which SP contributes to the enhancement of PSC power conversion efficiency. The PSC front electrode, which is typically made of ITO, is known to be a poor electrode due to its non-negligible sheet resistance and its tendency to degrade over time when in contact with acidic layers such as PEDOT:PSS [32]. Although finding a suitable replacement for ITO with a similar optical transparency and better conductivity has been challenging, the discovery of extraordinary optical transmission (EOT) through arrays of nanoholes by Ebessen [32] opened up the way to new promising candidates based on a thin metallic layer exhibiting SPR. The intensity of the electric field at the boundary of the nanoapertures (when light is passing through) is at its maximum and this extraordinary phenomenon can be used to enhance the light absorption of the materials within the perimeter [32,33]. Reilly et al. [32] have demonstrated that 30-nm thick Ag film populated with nanoholes of 92 nm in diameter improves the incident photon-to-current conversion efficiency (IPCE) by 50% compared to 130-nm thick ITO film. It should be noted that the plain 30-nm thick Ag film had demonstrated 20% less IPCE in Reilly's experiment. This improvement in efficiency is thought to be the result of the portion of photons that have been coupled to the surface of the metal, but could not transmit into the far field, yielding an increase in the overall absorbance [32].

Enhancement of PSCs by the incorporation of SP is not limited only to metallic nanogratings or thin film of nanohole arrays. The SP can also be generated in metallic nanoparticles (NPs), and due to their small dimensions, NPs can be easily used within the active polymer layer [34,35]. The presence of metallic NPs in the active layer not only increases the light absorption via SP generation, but also contributes to light scattering, which can serve as a light trapping structure within the cell [35]. The diameter of the NPs and the absorption efficiency are directly correlated, and it was shown that the larger metallic NPs improve the light absorption of the active layers [35]. Vedraine et al. [35] used numerical calculations to show that the spacing between the incorporated NPs also affects the absorption. Smaller spacing reduces considerably the absorption due to the reflection occurring at the surface of the NPs, preventing the light from reaching the active polymer layer [35].

The NPs are known to induce a localized plasmonic resonance (LPR), whereas the nanohole arrays and nanogratings on thin metallic film induce SPR [34]. The incorporation of these two types of resonance with narrow but yet different resonance wavelengths can considerably enhance the efficiency of PSCs by extending the absorption bandwidth in the active layer [34]. In a PSC depicted in Figure 5.6 a metallic grating and NPs were used. The metallic grating is used to enhance the power conversion efficiency by inducing SPR, but also acts as a back reflector that satisfies the light trapping strategy. The LPR is achieved by the NPs within the active

FIGURE 5.6 Schematic of a dual plasmonic (LPR and SPR) configuration used in a PSC. (Adapted from Li, X. et al. 2012. *Advanced Materials* 24:3046–3052.)

layer. Device configurations similar to Figure 5.6 have shown power conversion efficiency as high as 8.79% for a single junction polymer-fullerene-based BHJ solar cell [34].

5.3.3 QUANTUM DOTS

Apart from plasmonic structures, other techniques such as quantum dots (QDs) have been used to enhance PSC efficiency. Depending on their sizes and dimensions, QDs can increase the absorption of light in the active layers and lead to improvement of the power conversion efficiency [36]. The photoactive polymer in PSC cells is unable to absorb the NIR portion of the spectrum; therefore, the effective spectral range is limited to the visible wavelengths. However, this effective range can be expanded with the help of QDs [36]. It was shown that the changes in size and geometry of QDs can affect the charge separation efficiency of these nanoparticles, and this unique property can be used to extend the range of spectral absorption of PSCs [36]. Inorganic QDs such as CdSe are good electron acceptors and are used to enhance the charge transport within the device [37]. These QDs can be incorporated to interlayers such as Cs_2CO_3 and MoO_3 to modify the work function and improve the power conversion efficiency [37]. It was also shown that the incorporation of a CdSe QD layer can prevent Cs from diffusing from the interlayer into the active layer. QDs such as CdSe, CdTe, and PbTe are commonly used in PSCs, but despite their positive optical effects, ideal multiple carrier generation, and their low cost, these materials are extremely toxic and hazardous, which prevent them from being used in devices for large-scale manufacturing [38]. Graphene QDs (GQDs) can be a suitable substitute for the aforementioned QDs since their bandgap and, therefore, their optical properties are inversely proportional to their size [38].

5.3.4 INVERTED STRUCTURE

BHJ PSCs have shown a great potential for reaching power conversion efficiency in the vicinity of 8% using a combination of SP structures in an inverted PSC [34]. The advantage of the inverted structures is a good ohmic contact that provides a better charge-carrier collection within the active layer. It also prevents the acidic interlayer from direct contact with the ITO [39]. In an inverted structure, the ITO cathode is modified by n-type metal oxides or carbonates such as titanium oxide or caesium carbonate [39]. In this configuration, the use of PEDOT:PSS and low work function metal cathode is avoided. The top electrode enables a vertical phase separation in the active layer and is self-encapsulated by air-stable metals [39]. The achieved power conversion efficiency reported by inverted devices is much higher than the conventional structures used for PSCs; consequently, power conversion efficiencies greater than 10% can potentially be accessible in the near future.

Recently, He et al. [39] reported an inverted PSC structure with an efficiency of 9.2%, using a poly[(9,9-bis(3'-(N,N-dimethylamino) propyl)-2,7-fluorene)-alt-2,7-(9,9–dioctylfluorene)] (PFN) as the ITO surface modifier and a blend of [6,6]-phenyl C71-butyric acid methyl ester ($PC_{71}BM$) and thieno[3,4-b]thiophene/benzodithiophene (PTB7), which is a replacement for conventional P3HT. Figure 5.7 depicts the PSC

FIGURE 5.7 Schematic of an inverted PSC configuration with a MoO$_3$ as an interlayer. (Adapted from He, Z. et al. 2012. *Nature Photonics* 6:591–595.)

configuration used by He et al. The reported fill factor and the short-circuit current density (J_{SC}) were 69.99% and 17.46 mA/cm^{-2}, respectively [39].

5.4 MANUFACTURING TECHNIQUES: FROM SMALL-SCALE TO LARGE-SCALE PRODUCTION

One of the advantages of PSC technology when compared to traditional silicon-based solar cells is the exceptionally low initial capital investment required in specialized equipment and installation, and subsequently lower fabrication costs. Production of silicon photovoltaics remains based on silicon and integrated circuit (IC) processes, despite being costly and energy intensive. Compared to PSCs, Si-PVs have much higher associated material costs and lower production volumes; they require ultra-clean environments, high-temperature processing, and vacuum-based systems. In contrast, PSC technology takes advantage of commercially available equipment and low-temperature processes; they have shown the potential for in-air manufacturing and the upscaling of small-scale fabrication methodologies into large scale to reduce fabrication and material costs as well as capital investment. This section reviews manufacturing techniques commonly used at the laboratory scale and introduces processes required for large-scale manufacturing of PSCs.

5.4.1 SPIN COATING, SCREEN PRINTING, SPRAY COATING, AND INK-JET PRINTING

Among research laboratories, spin coating remains a widely used technique to deposit thin polymer films over flat surfaces with an unmatched precision over the thickness. Spin coating relies on centrifugal force to spread the polymer solution uniformly over the substrate. Films as thin as tens of nanometers can be achieved by choosing the appropriate viscosity, contact angle, and spin rate.

Screen printing is a commonly used industrial technique for fast and inexpensive deposition of films over large areas. In screen printing, the polymer ink is poured onto the screen and forced through the openings of the mesh by means of the movement of a blade or squeegee, as shown in Figure 5.8. Screen printing requires ink with viscosity 10 to 100 times higher than any other coating technique. Although the fabrication of PSCs by screen printing has been demonstrated [40], the low viscosity of the polymer inks makes it very challenging to control the thickness of the active polymer at the submicron scale.

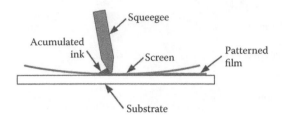

FIGURE 5.8   Schematic of a screen printing process.

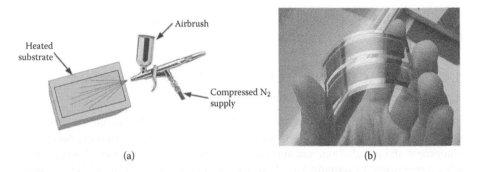

FIGURE 5.9   (a) Schematic of a spray coating setup, and (b) a PSC panel fabricated by spray coating at CiBER Labs (Simon Fraser University) using I|D|ME Technologies Corp. technology (US Patent Application 12/954614) of indium-based cathode for PSC stability improvement.

In spray coating, the polymer ink is atomized by the action of mixing the ink with compressed gas through the spray gun, creating small droplets of approximately one to tens of microns that are expelled toward the substrate, as shown in Figure 5.9(a). To obtain sufficient film morphology and uniformity, the droplets should dry upon impact with the surface. The thickness of the obtained film depends on several variables, including the viscosity of the polymer ink, the distance of the spray gun from the substrate, the surface roughness of the substrate, the surface temperature of the substrate, the flow rate in the spray gun cavity, and the compressed gas pressure. Although thin films can be obtained by spray coating, the resulting surface morphology is generally rough and prone to defects. Several research laboratories have demonstrated the effectiveness of this technique [41], including CiBER Labs (Simon Fraser University, Canada) with support provided by I|D|ME Technologies Corp. (US Patent Application 12/954614) in conjunction with an indium-based cathode for PSC stability improvement (see Figure 5.9b).

Industrial ink-jet printing is a relatively new technology with notable limitations in ink formulation and speed. The two basic methods for droplet formation and control are continuous (CIJ) and drop-on-demand (DoD) systems. CIJ printing is one of the earliest ink-jet technologies and operates in a continuous stream mode, making it a fast ink-jet method but lacking high-quality printing. In DoD systems, the droplets are created by the fluid displacement generated by the deformation of a piezoelectric

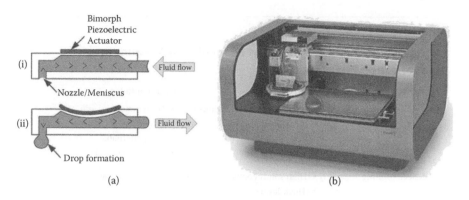

(a)   (b)

FIGURE 5.10 (a) Schematic of DoD operation (i: not actuated and ii: actuated). (b) A DoD ink-jet printer, Fujifilm Dimatix 2800 at CiBER Laboratory (Simon Fraser University, Canada).

actuator upon an applied electric pulse as depicted in Figure 5.10. By controlling the frequency and voltage, the droplet generation can be precisely adjusted, and the effectiveness of this technique has been demonstrated by several research groups [42,43]. Successful ink-jet printing of polymer solar cells requires the development of inks that dry sufficiently slowly, in combination with a printing scheme that fires narrowly spaced droplets to make single pixels coalesce over a reasonably large area before drying.

A summary of the advantages and disadvantages of each of the small and medium fabrication techniques is reported in Table 5.3.

5.4.2 ROLL-TO-ROLL PRINTING

Roll-to-roll (R2R) manufacturing is a process widely used to print polymer inks on flexible plastics or metal foils through a series of rolling cylinders at high speed in a large production volume. Typically, R2R processing refers to a combination of two or more of the techniques mentioned in this section.

Doctor blading is a coating technique commonly used in conjunction with R2R processes. In doctor blading, an excess of polymer ink is poured on the substrate, and a blade is used subsequently to disperse the polymer ink uniformly on the surface, as shown in Figure 5.11. Depending on the separation between the blade and the substrate, the desired thickness can be achieved. Other variables affecting the film formation are the viscosity of the ink, the rolling speed, and the temperature of the substrate.

Slot-die coating is another R2R-based printing technique where the ink is pumped through a coating die/applicator with defined slots. The dimension of the slot, the pump speed of the ink, the distance of the die from the substrate, and the rolling speed define the thickness of the wet layer, as shown in Figure 5.11(b). This coating technique has been demonstrated as a feasible suitable candidate for PSC fabrication [44].

TABLE 5.3

Advantages and Disadvantages of Small- and Medium-Scale Fabrication Processes

Fabrication Method	Advantages	Disadvantages
Spin coating	• Excellent thickness control • Homogeneous film • Low viscosity • Thin layers	• High material waste • Limited size • Not patternable • Flat surfaces only
Screen printing	• High solids content • Thick layers • Low-cost pattern change • Low printing pressure • Large-area application	• Poor resolution • Screen blocking • Low speed • Ink drying takes a long time
Spray coating	• Large-area application • Low-cost pattern change • Noncontact • Thin layers • Low-viscosity inks • Curved substrates possible	• Moderate waste • Generation of vapors • Relatively low resolution • Nonuniform films
Ink-jet printing	• High resolution • Noncontact • Thin layers • Low-viscosity inks • Low waste • Low-cost pattern change	• Stringent ink properties • Noncontinuous print • Limited ink reservoir

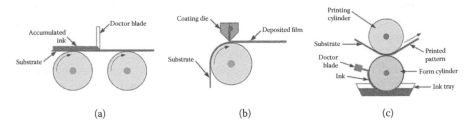

FIGURE 5.11 Schematic of (a) doctor blading coating, (b) slot-die coating, and (c) gravure printing.

A third technique commonly used for R2R manufacturing is known as gravure printing. Here a pattern is engraved in the form of very small cavities in a metallic cylinder; upon contact with a large reservoir of ink, the engraved cells on the cylinder are filled with a small volume ink to be applied to the substrate. A doctor blade is used to remove any excess of ink on the cylinder prior to application on the

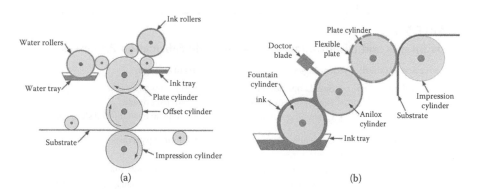

FIGURE 5.12 Schematic of (a) offset lithography printing, and (b) flexographic printing

substrate. During the imprinting step, the polymer ink is transferred from the cells to the substrate by a combination of surface tension and pressure applied on the printing cylinder as illustrated in Figure 5.11(c). The successful use of gravure printing for fabrication of PSCs has been reported previously [45,46].

Other large-area, high-volume fabrication techniques include offset lithography and flexographic printing, both illustrated in Figure 5.12. In offset lithography, surface wettability is used to control precisely the location of the polymer ink on the substrate. On the other hand, flexographic printing works in a similar way to gravure printing, with the exception that the flexible plate is mounted to a cylinder surface and is typically made from rubber (Table 5.4).

Roll-to-roll-based processes have a number of challenges that must be addressed before industrial-scale manufacturing of PSCs can be successful. For instance, the deposition of metal cathode requires physical vapor deposition and a highly controlled environment to prevent contamination of the interface between the cathode and the active layer, as well as potential oxidation of the metal. Silver-based pastes have been proposed as an alternative due to their ease of integration into the roll-to-roll process; however, conductive silver inks are a relatively new technology, and many silver-based inks require curing temperatures above the annealing temperatures of the polymer inks. Indium-based cathodes [17], for instance, can be deposited using a heat press or a slot die, which allows an easier integration into roll-to-roll processes. However, this technique has not yet been shown beyond the laboratory scale.

Nano-optics integration for industrial-scale fabrication processes presents even more challenges. For example, conventional direct writing nanofabrication tools such as focused ion beam or electron beam lithography are not compatible with standard R2R techniques and, in general, exhibit very slow throughput. However, recent work has shown that these state-of-the-art tools can be used to fabricate robust metal stamps that may be integrated into R2R processes via UV casting or hot embossing techniques [30].

TABLE 5.4
Advantages and Disadvantages of Large-Scale Fabrication Processes

Fabrication Method	Advantages	Disadvantages
Doctor blading	• Simple operation principle • Homogeneous films • Continuous coating	• High-viscosity inks • Not patternable • Frequent cleaning and sharpening
Slot die coating	• Continuous coating • Contact free • Low material waste • Precise ink application	• Limited patterns • Die clogging • High-viscosity inks • Complex cleaning and die change
Gravure printing	• Simple operation principle • High resolution • Low-viscosity inks • High speed and throughput • Excellent solvent resistance • Precise ink application	• Expensive printing cylinders • Solid tones reproduced via ink spreading • Missing dots • Smooth and compressible substrate
Offset lithography printing	• High resolution and speed • Low-cost printing plates • Versatile printing process • No excessive spreading • Waterless process possible	• Ink–water balance • Water limits the choice of materials • High surface strength of the substrate • Rough ink layer • Complex process • High pressure
Flexographic printing	• Accurate and adjustable ink transfer • Low-cost printing plates • Wide range of substrates • Low nip pressure	• Poor print resolution and register • Low printing speed • Difficult to print solid tones and details via the same printing unit • Poor solvent resistance of the plates

5.5 MANUFACTURING COST ANALYSIS FOR POLYMER SOLAR CELLS

Successful commercialization of an innovative product requires early viability assessment [47]. The investment methodology for new materials (IMM), established by Maine and Ashby [47], offers a guide on assessing research and product development directions. The first step to evaluating a technology is its viability, which involves assessing the technical performance, production cost, and market value. While the technical performance has been discussed in earlier sections, this section focuses on understanding the manufacturing cost of polymer solar panels and thereby assessing the viability of the technology.

A manufacturing cost analysis is essential to evaluating cost efficiency of manufacturing polymer solar panels and determining their price competitiveness against conventional silicon solar panels. The technical economic cost model is a cost analysis methodology developed to model cost structures more effectively by relying less on historical data, past experiences, and specific accounting practices [48]. In contrast, a technical economic cost model accounts for how changes in process control variables influence production rates and how changes in output volume affect relative cost of capital equipment, labor, raw materials, etc. Over the past 20 years, technical economic cost models (sometimes referred to as process-based models) have been utilized to compare manufacturing processes for novel technologies. These analyses are critical since manufacturing cost is a key element in considering an addition to new or improved performance attributes in any technology. For example, a manufacturing process may only be cost competitive on a low annual production volume, while losing this advantage at larger annual production volumes.

In general, a technical economic cost model consists of the following components:

1. Process input: working hours per day, working days per year, production rate, machine downtimes, wage rate, and product lifetime
2. Variable costs: raw materials, direct labor, and energy
3. Fixed costs: equipment purchase, manufacturing facility rental, fixed overhead, maintenance, tooling, working capital

Conducting a manufacturing cost analysis for polymer solar cells allows identification of an economically feasible process for scaled-up manufacturing. This analysis involves defining a manufacturing process for each of the following three types of production methods: batch, semicontinuous, and continuous. A batch production process requires less capital investment, but yields a smaller maximum annual production volume. A semicontinuous process requires more capital investment in automated production equipment, but yields a larger maximum annual output. Adoption of a continuous process requires a huge capital investment in fully automated equipment for roll-to-roll production, but the maximum annual output is also at larger scale.

Two major cost drivers for manufacturing scale-up are raw materials and direct labor, while other cost components in the technical economic cost model are less significant. The cost breakdown for the three production methods is as follows:

	Batch	Semicontinuous	Continuous
Raw materials	49%–75%	71%–94%	68%–97%
Direct labor	0.04%–17%	0.1%–20%	0.1%–22%
Others	Less significant	Less significant	Less significant
Total	100%	100%	100%

In all three manufacturing processes, it is evident that the cost of raw materials is the largest cost driver among all cost components. In the batch production process, cost of raw materials represents 75% of total production cost at small production volumes (i.e., 3 MW) annually. As production volume increases to 30 MW and then 300 MW annually, cost of raw materials reduces to only 49% with the batch process. In the semicontinuous process, cost of raw materials ranges from 94% in smaller production volumes to 71% in larger production volumes. Finally, in the continuous process, cost of raw materials ranges from 68% to 97% of total production cost, depending on the level of production. Direct labor cost represents a distant second important factor in determining manufacturing cost.

The manufacturing process of choice would depend in large part on the production demand of the entity and how much capital is available for investment in manufacturing equipment. Recent technological advances have warranted continued research to improve performances of polymer solar panels. The attributes that are key to improving overall performance include higher power conversion efficiencies and increased lifetime of the polymer solar cells. For instance, CiBER Labs (Simon Fraser University) is conducting research on integrating nanoscale structures into PSCs to potentially enhance their performances. Higher power conversion efficiencies will increase energy production capacity of a manufacturing process, thereby reducing the unit manufacturing cost of power produced.

In addition to technical performances, the price of raw materials such as the polymer inks/powders used in the functional layers of polymer solar cells is also crucial to the unit manufacturing cost. With improvements in technical performance, the manufacturing cost of polymer solar panels is expected to decrease in the coming years. By assuming that all cost components remain the same, the authors are working toward an improvement of power conversion efficiency to 60 W/m^2 and lifetime to 2 years, which would ultimately match the cost of conventional silicon solar panels. While they are still in early stages of development for large-scale, low-cost manufacturing, polymer solar technology is potentially a very low-cost option for flexible renewable energy due to its low material usage, low processing cost, and low installation cost.

REFERENCES

1. Barnes, D., Martinot, E., McCrone, A., Roussell, J., L. Sawin, J., Sims, R., et al. 2011. *Renewables 2011 Global Status Report.*
2. Marzencki, M., Tavakolian, K., Chuo, Y., Hung, B., Lin, P., and Kaminska, B. 2010. Miniature wearable wireless real-time health and activity monitoring system with optimized power consumption. *Journal of Medical and Biological Engineering* 30:227–235.
3. Xiaodan, Z., Xiaoyuan, X., Libin, Y., and Yong, L. 2009. A 1-V 450-nW fully integrated programmable biomedical sensor interface chip. *IEEE Journal of Solid-State Circuits* 44:1067–1077.
4. Park, S., Roy, A., Beaupre, S., Cho, S., Coates, N., Moon, J., et al. 2009. Bulk heterojunction solar cells with internal quantum efficiency approaching 100%. *Nature Photonics* 3:297–303.
5. Kalowekamo, J., and Baker, E. 2009. Estimating the manufacturing cost of purely organic solar cells. *Solar Energy* 83:1224–1231.
6. Wee, J. 2007. A feasibility study on direct methanol fuel cells for laptop computers based on cost comparison with lithium-ion batteries. *Journal of Power Sources* 173:424–436.
7. Gheidari, A. M., Behafarid, F., Kavei, G., and Kazemad, M. 2007. Effect of sputtering pressure and annealing temperature on the properties of indium tin oxide thin films. *Materials Science and Engineering: B* 136:37–40.
8. Jørgensen, M., Norrman, K., and Krebs, F. C. 2008. Stability/degradation of polymer solar cells. *Solar Energy Materials and Solar Cells* 92:682–714.

9. Rowell, M. W., Topinka, M. A., McGehee, M. D., Prall, H.-P., Dennler, G., Sariciftci, N. S., et al. 2006. Organic solar cells with carbon nanotube network electrodes. *Applied Physics Letters* 88:233506.

10. Pasquier, A. D., Unalan, H. E., Kanwal, A., Miller, S., and Chhowalla, M. 2005. Conducting and transparent single-wall carbon nanotube electrodes for polymer-fullerene solar cells. *Applied Physics Letters* 87:203511.

11. Na, S.-I., Kim, S.-S., Jo, J., and Kim, D.-Y. 2008. Efficient and flexible ITO-free organic solar cells using highly conductive polymer anodes. *Advanced Materials* 20:4061–4067.

12. Bundgaard, E. 2007. Low band gap polymers for organic photovoltaics. *Solar Energy Materials and Solar Cells* 91:954–985.

13. Coakley, K. M. 2004. Conjugated polymer photovoltaic cells. *Chemsitry of Materials* 16:4533–4542.

14. Scharber, M. 2006. Design rules for donors in bulk–heterojunction solar cells. *Advanced Materials* 18:789–794.

15. Troshin, P. A. 2008. Organic solar cells: Structure, materials, critical characteristics, and outlook. *Nanotechnologies in Russia* 3:242–271.

16. Schmidt-Mende, L. 2001. Self-organized discotic liquid crystals for high-efficiency organic photovoltaics. *Science* 293:1119–1122.

17. Omrane, B., Landrock, C., Chuo, Y., Hohertz, D., Aristizabal, J., Kaminska, B., et al. 2011. Long-lasting flexible organic solar cells stored and tested entirely in air. *Applied Physics Letters* 99:263305.

18. Bravo-Abad, J., Degiron, A., Przybilla, F., Genet, C., Garcia-Vidal, F. J., Martin-Moreno, L., et al. 2006. How light emerges from an illuminated array of subwavelength holes. *Nature Physics* 2:120–123.

19. Reilly, T. H., van de Lagemaat, J., Tenent, R. C., Morfa, A. J., and Rowlen, K. L. 2008. Surface-plasmon enhanced transparent electrodes in organic photovoltaics. *Applied Physics Letters* 92:243304.

20. Morfa, A. J., Rowlen, K. L., Reilly, T. H., Romero, M. J., and van de Lagemaat, J. 2008. Plasmon-enhanced solar energy conversion in organic bulk heterojunction photovoltaics. *Applied Physics Letters* 92:013504.

21. Ferry, V. E., Sweatlock, L. A., Pacifici, D., and Atwater, H. A. 2008. Plasmonic nanostructure design for efficient light coupling into solar cells. *Nano Letters* 8:4391–4397.

22. Lindquist, N. C., Luhman, W. A., Oh, S.-H., and Holmes, R. J. 2008. Plasmonic nanocavity arrays for enhanced efficiency in organic photovoltaic cells. *Applied Physics Letters* 93:123308.

23. Han, S. E., and Chen, G. 2010. Optical absorption enhancement in silicon nanohole arrays for solar photovoltaics. *Nano Letters* 10:1012–1015.

24. Niggemann, M., Riede, M., Gombert, A., and Leo, K. 2008. Light trapping in organic solar cells. *Physica Status Solidi* 511:2862–2874.

25. Duche, D., Drouard, E., Simon, J., Escoubas, L., Torchio, P., LeRouzo, J., et al. 2011. Light harvesting inorganic solar cells. *Solar Energy Materials and Solar Cells* 95:18–25.

26. Dal Zilio, S., Tvingstedt, K., Inganäs, O., and Tormen, M. 2009. Fabrication of a light trapping system for organic solar cells. *Microelectronic Engineering* 86:1150–1154.

27. Niggemann, M., Glatthaar, M., Gombert, A., Hinsch, A., and Wittwer, V. 2004. Diffraction gratings and buried nano-electrodes—Architectures for organic solar cells. *Thin Solid Films* 451-452:619–623.

28. Niggemann, M., Glatthaar, M., Lewer, P., Müller, C., and Wagn, J. 2006. Functional microprism substrate for organic solar cells. *Thin Solid Films* 511:628–633.

29. Zhu, X., Choy, W., Xie, F., Duan, C., Wang, C., He, W., et al. 2012. A study of optical properties enhancement in low-bandgap polymer solar cells with embedded PEDOT:PSS gratings. *Solar Energy Materials and Solar Cells* 99:327–332.

30. Chuo, Y., Landrock, C., Omrane, B., Hohertz, D., Grayli, S. V., Kavanagh, K., et al. 2013. Rapid fabrication of nano-structured quartz stamps. *Nanotechnology* 24:055304.
31. Baba, A., Aoki, N., Shinbo, K., Kato, K., and Kaneko, F. 2011. Grating-coupled surface plasmon enhanced short-circuit currentin organic thin-film photovoltaic cells. *ACS Applied Materials and Interfaces* 3:2080–2084.
32. Reilly, T. H., Lagemaat, J. V., Tenent, R. C., Morfa, A. J., Rowlen, K. L., and Reilly, T. H. 2008. Surface-plasmon enhanced transparent electrodes in organic photovoltaics. *Applied Physics Letters* 92:243304–243307.
33. Ebbesen, T. W., Lezec, H. J., Ghaemi, H. F., Thio, T., and Wolff, P. A. 1998. Extraordinary optical transmission through subwavelength hole arrays. *Nature* 391:667–669.
34. Li, X., Choy, W. C. H., Huo, L., Xie, F., Sha, W. E. I., Ding, B., et al. (2012). Dual plasmonic nanostructures for high performance inverted organic solar cells. *Advanced Materials* 24:3046–3052.
35. Vedraine, S., Torchio, P., Duche, D., Flory, F., Simon, J.-J., Le Rouzo, J., et al. 2011. Intrinsic absorption of plasmonic structures for organic solar cells. *Solar Energy Materials and Solar Cells* 95:57–64.
36. ten Cate, S., Schins, J. M., and Siebbeles, L. D. 2012. Origin of low sensitizing efficiency of quantum dots in organic solar cells. *ACS Nano* 6:8983–8988.
37. Lee, Y.-I., Youn, J.-H., Ryu, M.-S., Kim, J., and Jang, J. 2012. CdSe quantum dot cathode buffer for inverted organic bulk hetero-junction solar cells. *Organic Electronics* 13:1302–1307.
38. Gupta, V., Chaudhary, N., Srivastava, R., Sharma, G. D., Bhardwaj, R., and Chan, S. 2011. Luminscent graphene quantum dots for organic photovoltaic devices. *Journal of the American Chemical Society* 133:9960–9963.
39. He, Z., Zhong, C., Su, S., Xu, M., Wu, H., and Cao, Y. (2012). Enhanced power-conversion efficiency in polymer solar cells using an inverted device structure. *Nature Photonics* 6:591–595.
40. Krebs, F. C., Jørgensen, M., Norrman, K., Hagemann, O., Alstrup, J., Nielsen, T. D., et al. 2009. A complete process for production of flexible large area polymer solar cells entirely using screen printing—First public demonstration. *Solar Energy Materials and Solar Cells* 93:422–441.
41. Girotto, C., Rand, B. P., Genoe, J., and Heremans, P. 2009. Exploring spray coating as a deposition technique for the fabrication of solution-processed solar cells. *Solar Energy Materials and Solar Cells* 93:454–458.
42. Aernouts, T., Aleksandrov, T., Girotto, C., Genoe, J., and Poortmans, J. 2008. Polymer-based organic solar cells using ink-jet printed active layers. *Applied Physics Letters* 92:12–13.
43. Eoma, S. H., Park, H., Mujawar, S., Yoon, S. C., Kim, S., Na, S., et al. 2010. High efficiency polymer solar cells via sequential ink-jet-printing of PEDOT:PSS and P3HT:PCBM inks with additives. *Organic Electronics* 11:1516–1522.
44. Krebs, F. 2009. All solution roll-to-roll processed polymer solar cells free from indium-tin-oxide and vacuum coating steps. *Organic Electronics* 10:761–769.
45. Kopola, P., Aernouts, T., Guillerez, S., Jin, H., Tuomikoski, M., Maaninen, A., et al. 2010. High efficient plastic solar cells fabricated with a high-throughput gravure printing method. *Solar Energy Materials and Solar Cells* 94:1673–1680.
46. Ding, J. M., Vornbrock, A., Ting, C., and Subramanian, V. 2009. Patternable polymer bulk heterojunction photovoltaic cells on plastic by rotogravure printing. *Solar Energy Materials and Solar Cells* 93:459–464.
47. Maine, E. M. A., and Ashby, M. F. 2002. An investment methodology for new materials. *Materials and Design* 23:297–306.
48. Clark. J, R. R. 1997. Techno-economic issues in material science. In *ASM Handbook Materials* Park, OH, 20.

6 Inverted Organic Solar Cells

Purna P. Maharjan and Qiquan Qiao

CONTENTS

6.1 INTRODUCTION

Polymer solar cells (PSCs), a potential green energy source, are under intense investigation in both academic and industrial sectors [1-12]. To overcome today's energy issues, utilization of the ubiquitous solar energy has become a promising idea worldwide [13]. At present, inorganic photovoltaics (PVs) are the dominant technology in the market. However, high material and manufacturing costs limit their wide acceptance. Organic photovoltaics (OPVs) provide promise as a low-cost technology due to their advantages of light weight, mechanical flexibility, and mass production by roll-to-roll solution processing. Significant progress has also been made in the production of organic light-emitting devices (OLEDs) and realization of thin-film transistors (TFTs) [14,15]. Intense research has been carried out to obtain high-efficiency, stable, low-cost, and high-speed production of organic solar cells [16]. Current research on synthesis of new narrow bandgap materials to absorb a broader range of the solar spectrum [17], optimization of phase segregation in the

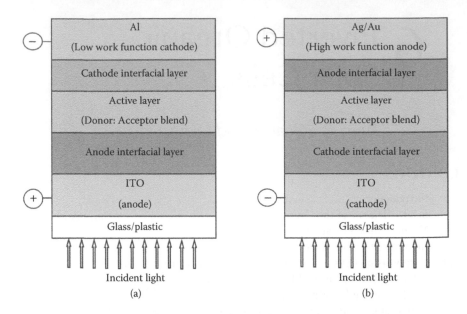

FIGURE 6.1   (a) Normal and (b) inverted structure polymer solar cell configuration.

bulk heterojunction (BHJ) layers [18], interfacial modification for better charge carrier collection [19,20], design of novel configuration, and improvement in processing [18] have dramatically increased the efficiency and lifetime in recent years.

Regular device structure, as shown in Figure 6.1(a), with low work function metal as the top electrode is generally used to fabricate polymer solar cells [21,22]. Work functions of electrodes (anode and cathode) should be aligned with the highest occupied molecular orbits (HOMOs) of donors and the lowest unoccupied molecular orbits (LUMOs) of acceptors for effective charge collection. For this purpose, low work function metals have been chosen as cathodes for effective matching with the LUMO of the acceptor [23,24]. With regular device structure, power conversion efficiencies (PCEs) for bulk heterojunction PSCs have been over 8% [25]. However, this device structure has some disadvantages:

1. Need of a high vacuum to deposit the top metal electrode, which is not suitable for mass production by roll-to-roll processing [26]
2. Low lifetime due to degradation of cells in air
3. Inherent vertical phase segregation in the polymer active layer [27,28]

To overcome these limitations, inverted configuration has been used widely in recent years; this involves high work function metal or metal oxide as the top electrode (anode) and low work function metal as the bottom electrode (cathode), as shown in Figure 6.1(b). This structure reverses the nature of charge collection as compared to regular structure devices. This higher work function top electrode (metal oxides and metal) protects cells from damage by moisture and oxygen in air, enabling the ideal roll-to-roll processing and high stability [28,29]. High and low work function electrodes as anode and cathode that are commonly used are silver (Ag) or gold (Au) and indium tin oxide, (ITO), respectively [30–33]. Furthermore, utilization of the inherent vertical phase separation in the active layer enhances the efficiency [27]. Various n-type transparent oxides including ZnO, TiO_2/TiO_x, Cs_2CO_3, and Nb_2O_5 are used as an interfacial layer to collect electrons efficiently [29,33–38]. Conjugated semiconducting electrolytes and self-assembled functional molecules have recently drawn attention as surface modifications of ITO and even replacing the transparent conducting oxides for inverted device structure [19,20]. A very thin layer of these materials is deposited to ensure high transparency so that sufficient light gets absorbed by the active layer. Similarly, p-type oxides such as NiO_x, VO_x, WO_x, and MO_x are used between the active layer and the top metal as anodic modification to collect holes efficiently [39–44].

Effectiveness of these materials also depends on the electrode work function and active layer materials. So selection of interfacial layers should support an ohmic contact between electrodes and active material. In 2006, Li et al. [31] demonstrated that relative positioning of two interfacial buffer layers; that is, vanadium oxide (V_2O_5) and cesium carbonate (Cs_2CO_3) can alter the polarity of a solar cell. V_2O_5 was used as the hole injection layer and Cs_2CO_3 as an electron injection layer. Utilizing V_2O_5 and Cs_2CO_3 interfacial layers, both regular and inverted polymer solar cells can be made [31]. In 2009, Krebs, Gevorgyan, and Alstrup [45] demonstrated inverted device geometry with substrate/ITO/ZnO/poly(3-hexylthiophene) (P3HT):fullerene derivative [6,6]-phenyl C61 butyric acid methyl ester (PCBM)/poly(3,4-ethylene dioxythiophene doped with polystyrene sulfonate) (PEDOT:PSS)/Ag and compared its performance with a regular structure device of substrate/ITO/PEDOT:PSS/P3HT:PCBM/Al. For both structures, power conversion efficiency reached 2.7% under ambient conditions (AM1.5G, 1000 Wm^{-2}, 40% ± 5% relative humidity). Furthermore, they transferred the inverted device model to a full roll-to-roll process. A flexible substrate, polyethyleneterephthalate (PET), was coated with ITO. After converting the ITO layer into stripes, the ZnO layer, active layer, and PEDOT:PSS layer were deposited using a slot-die coating. Then silver paste was coated on the top using a screen printing technique [45].

6.1.1 THEORY OF THE ORGANIC SOLAR CELL

OPV devices are based on carbon-based organic semiconductors in which charge transport is mainly by the delocalization of electrons along conjugated polymer backbones [16]. Carbon atoms are sp^2 hybridized and form three sigma bonds with neighboring atoms. The fourth orbital (P_z) is perpendicular to the plane of sp^2 orbitals. Delocalization of the charges in organic semiconductors is due to the overlapping of the P_z electron wave functions [46]. Upon light illumination, photons are absorbed by an active layer and electron–hole pairs are generated as excitons. Unlike inorganic semiconductors, organic semiconductors have a relatively low dielectric constant ($\varepsilon_r \approx 2$–4). This causes the coulombic interaction between electron and hole to be strong and form tightly bonded excitons. Typical binding energy of Frenkel exciton is around 0.3–1.0 eV [13,47].

Effective conversion of light into electricity depends on several steps: (1) absorption of photons to create excitons, (2) diffusion of excitons to interface, (3) dissociation of excitons into free charges, and (4) charge transport to respective electrodes. The absorption efficiency (η_A) depends on absorption spectra of organic materials, film thickness, and device architecture [48]. For donor–acceptor-based devices, light is mainly absorbed in donor material (i.e., a conjugated polymer). Since the absorption coefficient of organic materials is high (typically exceeding 10^5 cm^{-1} [49]), a very low thickness of 100–300 nm is sufficient for efficient light absorption. Improved absorption efficiency can be achieved by utilizing low bandgap materials [50]. In addition, tandem device structures combining two or more subcells with complementary absorption bands can be used to harvest a broader spectrum of solar radiation [51].

Once excitons are generated, they diffuse toward the interface. The exciton diffusion efficiency (η_{ED}) mainly depends on the exciton diffusion length (L_D) and donor–acceptor interface. Typical exciton diffusion length in conjugated polymers is ~4–20 nm [52,53]. If the distance between exciton generation location and donor–acceptor interface (L_i) is larger than L_D, then excitons will decay through radiative or nonradiative recombination before reaching to the interface. L_D is equal to $(D\tau)^{1/2}$, where D is diffusion coefficient and τ is exciton lifetime [54]. Strong electric fields are needed to dissociate the excitons, which can be supplied through externally applied electric fields or through interfaces. Since, at interface, change of potential energy is tremendous, it acts as a local strong electrical field for exciton dissociation [53]. The exciton dissociation (or charge transfer) efficiency (η_{CT}) depends on the donor–acceptor LUMO offset and/or internal electric field at donor–acceptor interface, which should exceed exciton binding energy [55,56]. Photophysics study of conjugated polymers in solar cells has shown that photo-induced charge transfer from donor to acceptor is ultrafast within 100 fs [57]. Similarly, the charge collection efficiency (η_{CC}) depends on the morphology and mobility of the active layer. The mobility has an impact on carrier extraction and losses through recombination [58]. Thus, the overall efficiency of current generation under incident light (i.e., external quantum efficiency (η_{EQE})) can be calculated as $\eta_{EQE}(\lambda,V) = \eta_A(\lambda).\eta_{ED}.\eta_{CT}(V).\eta_{CC}(V)$, where V is the voltage across the cell and λ is the wavelength of incident light [48]. In addition, other parameters that determine the cell efficiency are short circuit current (J_{sc}), open circuit voltage (V_{oc}), fill factor (ff), series resistance (R_s), and shunt resistance (R_{sh}) [24].

6.2 NORMAL STRUCTURE SOLAR CELLS

At early stages, most organic solar cells, including single layer, bilayer, and bulk heterojunction devices, were fabricated using normal device structure by sandwiching photoactive layers between two electrodes. Generally, higher work function electrode ITO served as the anode and the low work function electrode Al as cathode [59]. For example, normal device structure of ITO/PEDOT:PSS/polymer:PCBM/Al uses ITO to collect holes and Al to collect electrons. The PCEs of single layer solar cells with pure conjugated polymers were very small, in a range of $10^{-3}\%$ to $10^{-1}\%$ [57]. In 1986, Tang demonstrated a two-layer organic solar cell using two conjugated small molecules: copper phthalocyanine (CuPc) and a perylene tetracarboxylic derivative [60]. Although the power conversion efficiency was low (around 1%), it was a milestone for the development of organic solar cells using donor–acceptor blends. Much work was then done to improve the performance of the bilayer organic solar cells.

However, efficiency of organic solar cells with both a single layer and bilayers was limited by exciton diffusion length and the area of donor–acceptor interface. The planar junction concept provides a very small interfacial area of donor–acceptor interface, limiting dissociation of excitons and hence photocurrent generation [61]. This problem was drastically solved by the introduction of a new concept known as "bulk heterojunction." This concept includes mixing of donor and acceptor materials so that both interpenetrate each other, forming large donor–acceptor interfacial area for exciton dissociation. In addition, it provides the nanoscale bicontinuous percolated pathways for charge transport [46,57,61]. The BHJ structure showed a great advantage over single layer or bilayer structures as excitons can dissociate effectively throughout the active layer, generating comparatively large numbers of free electrons and holes. It was first demonstrated by Hiramoto, Fujiwara, and Yokoyama via coevaporation of donor and acceptor molecules under high vacuum [62,63]. However, the first efficient bulk heterojunction polymer solar cells were shown by researchers in 1995 using polymer–polymer and polymer–fullerene blends [61,64]. Fullerene has become the dominating electron acceptor as it has high electron affinity and capability of charge transport [57]. Regular solar cells have achieved efficiency up to 8% [25]. However, some issues, including vertical phase separation and stability, exist in regular structure solar cells. A new device architecture called inverted bulk heterojunction (IBHJ) structure has recently emerged as a dominating structure for longer stability and higher device performance.

6.3 INVERTED STRUCTURE SOLAR CELLS

As discussed earlier, inverted structure device uses high and low work function electrodes as anode and cathode, respectively, reversing the nature of charge collection compared to normal device configuration. Ag/Au and ITO are used as anode and cathode, respectively. Generally, plastic or glass coated with ITO is used as substrate. ITO is widely used as an electrode because of its high conductivity and transmittance, and suitable work function. Since ITO is expensive, several alternatives, such as graphene, aluminum doped ZnO [65], carbon nanotubes, and conducting polymer films have also been used [66–68]. Some interfacial layers, including Cs_2CO_3, Ca, ZnO, TiO_2, AZO, Nb_2O_5, and nonconjugated polyelectrolytes (NPEs), have been used to modify ITO for effective collection of electrons [19,20,29,33–38].

These layers are normally very thin to ensure adequate transparency for light to reach the active layer. Most of these layers are solution processed and spin coated on the top of the ITO surface, which provides less fabrication complexity. Polymers are typically spin coated using a solution process because they have large molar mass and can decompose under excessive heating. Donor–acceptor materials are dissolved in a solvent such as dichlorobenzene (DCB), chlorobenzene (CB), chloroform, and toluene [53,69–71]. Some other techniques, such as doctor blading, screen printing, and ink-jet printing, can also be used for active layer deposition. For small molecules, vacuum evaporation/sublimation are usually used to deposit thin film. Usually, a vacuum $< 10^{-5}$ mbar is maintained to deposit thin films by thermal evaporation [53,72]. Similarly, some hole transporting layers such as MoO_3, WO_3, NiO, V_2O_5, and PEDOT:PSS have been used between the anode and active layer for effective collection of holes [39–44]. PEDOT:PSS is deposited by solution processing, but MoO_3, V_2O_5, and WO_3 are generally deposited by evaporation under high vacuum. However, some people have also reported solution processing of MoO_3, V_2O_5, and WO_3 [73–75]. Device fabrication using solution processing techniques of all involved layers is the ultimate goal of organic solar cells so that roll-to-roll processing and mass production are viable, which can reduce the cost to a great extent.

For example, an inverted device structure of ITO/ZnO/polymer:PCBM/MoO_3/ Ag utilizes ITO to collect electrons and Ag to collect holes. ZnO transports electrons from PCBM to ITO (cathode) and MoO_3 transports holes from polymer to Ag (anode). In 2005, Glatthaar et al. [76] reported a device configuration with inverted layer sequence and compared cell performance with a regular structure device. Al was chosen as the bottom electrode and deposited on the top of glass followed by electron-beam deposition of a titanium layer (20 nm). Then P3HT:PCBM was spin coated on the titanium layer. The PEDOT:PSS layer was placed between the photo-active layer and the Au grid (anode). Light was illuminated from the Au grid side and an efficiency of 1.4% ± 0.3% was obtained under AM 1.5 solar illumination [76]. In 2008, the Yang group [37] demonstrated an inverted device structure with glass/ ITO/Cs_2CO_3/RR-P3HT:PCBM/V_2O_5/Al. They reported the inverted device structure with an efficiency from 2.3% to 4.2%. Further, they demonstrated that device efficiency can be improved by reducing the work function with annealing treatment. X-ray photoelectron spectroscopy (XPS) results showed that the modification of Cs_2CO_3 work function can be tuned from 3.45 to 3.06 eV with an annealing treatment at temperatures below 200°C. They used XPS to reveal that Cs_2CO_3 decomposes into an n-type doped semiconductor after the thermal annealing process [37].

In 2012, He et al. [77] demonstrated a certified efficiency of 9.2% with an inverted device structure. An alcohol-/water-soluble conjugated polymer, poly [(9,9-bis(3′- (N,N-dimethylamino)propyl)-2,7-fluorene)-alt-2,7-(9,9–dioctylfluorene)] (PFN) was used to modify ITO surface. XPS measurements showed that deposition of a thin layer of PFN (10 nm) reduces the ITO work function from 4.7 to 4.1 eV. This modified ITO forms an ohmic contact with the photoactive layer and can be used as the cathode for an inverted device structure. The overall device structure for inverted configuration was ITO/PFN/PTB7:PC71BM/MoO_3/Al or Ag. Furthermore, by replacing the highly reflective anode MoO_3 (10 nm)/Al (100 nm) with a transparent layer of MoO_3 (10 nm)/Ag (20 nm), they also demonstrated a semitransparent polymer solar cell that can be applied in building windows, foldable curtains, and invisible circuits. Under front and rear side AM 1.5G illumination, they reported an efficiency of 6.13% and 4.96%, respectively [77].

6.4 COMPARISION BETWEEN INVERTED AND REGULAR STRUCTURES

6.4.1 Effect of Vertical Phase Separation

Spin coating is a common process for making uniform thin polymer–acceptor films. The polymer–acceptor blend is phase separated during the spin-coating process, forming a complex morphology. The blend of two polymers, including polystyrene (PS) and polymethylmethacrylate (PMMA), undergoes vertical stratification, making the interface between the stratified layers unstable; this finally forms the phase-separated thin film [78]. The vertical phase segregation of the polymer blend is due to the surface energy difference between the components of the blend. The component with lower surface energy is attracted toward the interface with air to lower the overall energy. Similarly, the component with higher surface energy is attracted toward the substrate [27,79]. Xu et al. [27] studied the vertical phase separation of P3HT:fullerene derivative blends. Through XPS and atomic force microscopy (AFM) analysis, they observed enrichment of P3HT at the free (air) surfaces and fullerene derivatives at the organic/substrate interfaces. They prepared a blend of PCBM and [6,6]-(4-fluoro-phenyl)-C61-butyric acid methyl ester (FPCBM) with P3HT separately and spin coated on glass. For XPS analysis, blend films were lifted off the substrates and transferred to a conductive surface.

Figure 6.2 represents the comparison of compositions of the films spin coated on glass substrates prepared by fast grow, slow grow, fast grow with annealing, and slow grow with annealing, respectively. The fast-grow and slow-grow films showed higher PCBM or FPCBM concentration on the bottom surfaces at the organic/substrate interface. When thermally annealed, the slow-grow films showed invariant concentration of PCBM or FPCBM, while fast grow showed slight decrease in PCBM of FPCBM concentrations at both sides of the films. The comparison based on XPS results indicated that this vertical phase separation of donor and acceptor can be beneficial for inverted device structure because inverted structure will involve

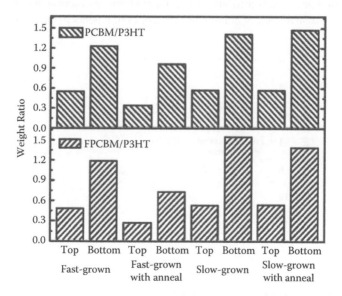

FIGURE 6.2 Compositions of the blend films at the top surface (free air) and bottom surface (organic/substrate interfaces) spin-coated on glass substrate. (Reproduced with permission from Z. Xu et al. 2009. *Advances in Functional Materials* 19:1227–1234.)

the polymer:fullerene blend film with acceptor-enriched regions toward the cathode and donor-enriched regions toward the anode. This alignment of regions toward corresponding electrodes can assist in efficient charge dissociation, charge transport, and charge collection at the interface. [27].

Subbiah et al. [80] studied the effect of vertical distribution of blended components in bulk heterojunction solar cell based on poly[(4,4′-bis(2-ethylhexyl)dithieno[3,2-b:2′,3′-d]silole)-2,6-diyl-alt-(2,1,3-benzothiadiazole)-4,7-diyl](PDTS-BTD) and [6,6]-phenyl-C71 butyric acid methyl ester (PC$_{71}$BM). XPS and Auger electron spectroscopy (AES) depth profiles were measured and they observed that PC$_{71}$BM was rich at the bottom surface and polymer-rich region on the top surface of PDTS-BTD/PC71BM blend films. To compare the effect of vertical phase segregation on device performance, they fabricated both regular and inverted solar cells. Regular device structure was glass/ITO/PEDOT:PSS (or MoO$_3$)/PDTS-BTD:PC$_{71}$BM/LiF/Al and inverted device structure was glass/ITO/ZnO/PDTS-BTD:PC$_{71}$BM/MoO$_3$/Ag. The illuminated J–V characteristics are shown in Figure 6.3; a PCE of 4.42% with a J_{sc} of 13.08 mA/cm^2, V_{oc} of 0.61 V, and an ff of 0.55 was achieved for regular device structure using PEDOT:PSS. Replacing PEDOT:PSS with MoO$_3$ in a regular device, a PCE of 4.91% with a J_{sc} of 12.97 mA/cm^2, V_{oc} of 0.62 V, and an ff of 0.61 was observed. However, the inverted device with the ZnO electron transport layer showed enhanced device performance compared to a regular device. A higher PCE of 5.51% with a J_{sc} of 16.48 mA/cm^2, V_{oc} of 0.58 V, and ff of 0.57 was demonstrated in the inverted device. In addition, inverted devices showed a maximum EQE of 63% at 720 nm compared with 50% in regular devices. This significant PCE enhancement in the inverted device was attributed to efficient charge extraction due to inhomogeneous vertical distribution of donor and acceptor components in the active layer [80].

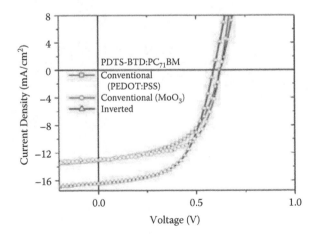

FIGURE 6.3 J–V characteristics of PDTS-BTD:PC71BM solar cells with regular and inverted architecture. (Reproduced with permission from J. Subbiah et al. 2012. *Solar Energy Materials and Solar Cells* 97:97–101.)

6.4.2 DEVICE STABILITY

Regular organic solar cells (OSCs) involve PEDOT:PSS on the top of ITO. De Jong, Van Ijzendoorn, and De Voigt [28] used Rutherford backscattering (RBS) to show that the ITO/PEDOT:PSS interface is unstable due to etching of ITO from the strong acidic nature of PEDOT:PSS. Degradation of the interface was observed as certain concentrations of indium migrated to the PEDOT:PSS film. Figure 6.4 shows the migration of indium into PEDOT:PSS during 0–20 h. They reported 0.02 at.% indium migration into the PEDOT:PSS film in prepared samples. An increase of indium concentration to 0.2% was found after 10 h exposure to air. When annealed in a nitrogen environment at 100°C, it took 2500 h to increase indium concentration to 0.2%. This indicates that the ITO/PEDOT:PSS layer is very sensitive to air. Due to the hygroscopic nature of PSS, PEDOT:PSS film can absorb moisture when exposed to air and forms an aqueous acid environment, which can etch ITO and cause the etched products to migrate into the PEDOT:PSS film [28].

Lee et al. [81] demonstrated a more stable and efficient inverted solar cell than regular structure solar cells. Inverted devices with ITO/Cs$_2$CO$_3$/active layer/MoO$_3$/Al showed a higher PCE of 4.01% compared to 3.63% in regular solar cells with ITO/PEDOT:PSS/active layer/LiF/Al. Furthermore, they compared the stability of inverted with regular devices in air. Figure 6.5 shows the stability of inverted and regular BHJ solar cells in air as a function of time. Within 40 days at 295 K, the PCE of the inverted solar cell decreased to about 93.5% of its initial value while the regular structure solar cell decreased to 87.7%. Absence of PEDOT:PSS and use of the MoO$_3$ layer for hole transport were the possible reason for improved stability in inverted over regular structure device. In addition, they also demonstrated that the optical field distribution in the active layer is better for an inverted device using the MoO$_3$ layer [81].

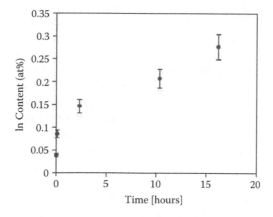

FIGURE 6.4 Indium content in PEDOT:PSS films as the time of exposure of the films to air. (Reproduced with permission from M. De Jong, M., L. Van Ijzendoorn, and M. De Voigt. 2000. *Applied Physics Letters* 77:2255–2257.)

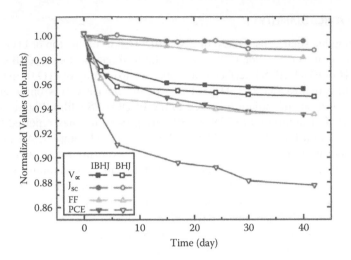

FIGURE 6.5 Plot of IBHJ and BHJ solar cell parameters normalized to initial values as the time of exposure in air. (Reproduced with permission from Y.-I. Lee et al. 2012. *Organic Electronics* 12:353–357.)

Chu et al. [82] demonstrated an inverted BHJ solar cell using annealing-free ZnO nanocrystals as an electron transport layer. An active layer was a blend of a low-bandgap polymer and $PC_{71}BM$ with nitrobenzene by 4% volume as an additive. A significant PCE of 6.7% was achieved with higher device stability. These unencapsulated inverted solar cells retained 85% of original efficiency, even after 32 days of storage in air. They demonstrated that processing additive makes an important contribution to device stability [82]. Use of ZnO in inverted PSCs can block the UV light and protect the active layer from photodegradation, resulting in longer device lifetime. A device with ZnO as buffer layer showed no degradation when continuously illuminated for 4 h at AM 1.5. In addition, the efficiency only dropped by 5% and 10% after 47 and 160 days of storage in a glove box respectively [25].

Hau et al. [29] compared the stability and performance of regular BHJ solar cells with inverted solar cells with ZnO as an electron selective layer. Regular devices with LiF/Al as electrode showed an average PCE of ~2.4% compared to those of ~3.5% and ~3.6% for inverted solar cells with sol-gel processed ZnO and ZnO nanoparticles respectively. The improvement of performance in inverted solar cells is contributed to the ZnO sol-gel layer, providing an additional n-type interface for photocurrent generation. Under storage in ambient conditions, regular devices with LiF/Al as electrodes degraded very fast, as PCE reduced below half of its original value after 1 day and even died fully after 4 days. However, PCE of the inverted device structure retained above 80% over 40 days in air. The improved stability is contributed to the PEDOT:PSS/Ag electrode as a cathode. PEDOT:PSS on top of the active layer in this case prevented the active layer from oxygen. Ag electrode upon oxidation in air can form silver oxide (AgO), modifying the work function from −4.3 to −5.0 eV. The work function of AgO is more aligned with that of PEDOT:PSS (−5.1 eV) for effective charge collection [83,84]. However, in regular BHJ solar cells where low work function metals such as Ca/Al are used as cathodes, Ca reacts with oxygen and water at room temperature and degrades devices [85].

Using experimental refractive indices and extinction coefficients with optical modeling based on one-dimensional transfer matrix formalism (TMF), He et al. [77] showed that an inverted structure device can harvest optimum light from the incident solar spectra. They also studied the stability of encapsulated normal and inverted structure devices under ambient conditions. After 62 days, the inverted structure device showed a higher stability, retaining 95% of its initial efficiency (~9%). However, normal device structure degraded to half of its original efficiency after 10 days. These results justify that inverted structure devices can provide higher performance and stability.

6.5 DIFFERENT CATHODE INTERFACIAL LAYERS USED IN INVERTED SOLAR CELLS

6.5.1 ZnO

In 2008, Kyaw et al. [33] reported an inverted solar cell with ZnO and MoO_3 as electron- and hole-selective layers, respectively. The molar concentration of ZnO prepared from zinc acetate precursor was varied from 0.75 to 0.1 M in order to increase the optical transmission from 75% for 0.75 M to 95% for 0.1 M. The inverted device architecture was FTO/ZnO/P3HT:PCBM/MoO_3/Ag. Because of the higher transmittance of 0.1 M ZnO over 0.75 M ZnO, the J_{sc} and ff improved from 5.986 mA/cm^2 and 47% to 8.858 mA/cm^2 and 57% [33]. Later in 2008, Hau et al. [86] investigated and compared the performance of both inverted and regular structure solar cells by employing low-temperature-processed ZnO nanoparticles as an electron selective layer. The ZnO buffer layer reduces oxygen diffusion into the photoactive layer and minimizes chemical or physical damage in the photoactive layer. ZnO exhibited high electron mobility and low-temperature-induced crystallinity. The regular device structure is glass/ITO/PEDOT:PSS/P3HT:PCBM/LiF/Al, while the inverted structure is glass/ITO/ZnO/P3HT:PCBM/PEDOT:PSS/Ag. The regular structure gave, on average, a J_{sc} = 7.87 mA/cm^2, V_{oc} = 0.598 V, ff = 50.9%, and PCE of ~2.4%; the inverted structure resulted in an average performance with J_{sc} = 11.08 mA/cm^2, V_{oc} = 0.617 V, ff = 51.3%, and PCE of ~3.5%. The ff was similar for both device structures, but the J_{sc} and V_{oc} improved in the inverted architecture. The enhanced performance in inverted solar cells was attributed to the additional n-type interface provided by the ZnO nanoparticles [86].

In 2010, Cheun et al. [36] reported an inverted polymer solar cell using a ZnO-modified ITO electrode as the electron selective electrode. The ZnO, with varying thickness from 0.1 to 100 nm, was deposited by atomic layer deposition (ALD). They studied the effect of ZnO thickness variation on the ITO/ZnO properties and device performance. The device structure used in their work was glass/ITO/ZnO/P3HT:PCBM/PEDOT:PSS/Ag. As illustrated in Figure 6.6(a), the J–V characteristics at ZnO thickness less than 10 nm showed an S-shaped curve resulting in low efficiency below 1%. However, no S-shape was observed for the ZnO layer with thickness greater than 10 nm, and an efficiency of 2.13% was achieved. Figure 6.6(b)

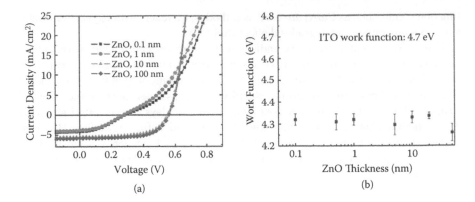

FIGURE 6.6 (a) The J–V curve of the P3HT:PCBM inverted solar cell incorporating ZnO electron transport layer and (b) the variation of ITO work function as a function of ZnO thicknesses. (Reproduced with permission from H. Cheun et al. 2010. *Journal of Physical Chemistry C* 114:20713–20718.)

illustrates the dependence of ITO/ZnO work function on the ZnO layer thickness. It was found that there was negligible variation of the ITO/ZnO layer work function on the ZnO thicknesses. While the ITO work function was measured at 4.7 eV, the work function of the ZnO-modified ITO was around 4.3 eV, which is close to the LUMO level of PCBM. From the AFM measurement, the surface roughness of 1.89, 2.32, and 2.41 nm for 5, 10, and 20 nm ZnO films, respectively, was also considered insignificant to cause the difference in the J–V characteristics. They explained that the thinner ZnO layer has a larger surface-to-bulk ratio, causing grain defect to play a significant role in reducing the electrical conductivity of the ZnO film. The reduced conductivity originates from the presence of oxygen at grain boundaries, which acts as an electron trap. However, this low conductivity can be improved by UV irradiation. UV illumination induces photogenerated electrons and holes in ZnO. These photogenerated holes desorb negatively charged oxygen molecules trapped at the grain boundaries, whereas remaining photogenerated electrons act as free electrons increasing the charge carrier density of ZnO [36,87,88].

In 2012, Ka et al. [89] investigated the effect of annealing temperature of ZnO prepared from an aqueous solution of ammine-hydroxo zinc complex on device performance of an inverted polymer solar cell. The spin-casted ZnO was annealed at temperatures ranging from 25°C to 150°C. For the device structure of glass/ITO/ZnO/P3HT:PCBM/-MoO$_3$/Ag (Figure 6.7), the ZnO annealed at temperatures ranging from 25°C to 60°C exhibited S-shaped J–V curves with PCE less than 0.2%. This S-shaped feature at lower temperatures was thought to result from low probability of dissociated excitons reaching the electrodes. However, a sudden change in J–V behavior was observed at 70°C annealing temperature, resulting in $J_{sc} = 9.7$ mA/cm^2, $V_{oc} = 0.59$ V, $ff = 0.49$, and PCE = 2.6%. At annealing temperature of 80°C and above, the device efficiency improved and saturated with $J_{sc} = 10.0$ mA/cm^2, $V_{oc} = 0.59$ V, $ff = 0.62$, and PCE = 3.6%. The values of the optical bandgap Eg obtained from transmission spectra reduced from 5.60 eV at 25°C to 3.45 eV at 80°C suggested that

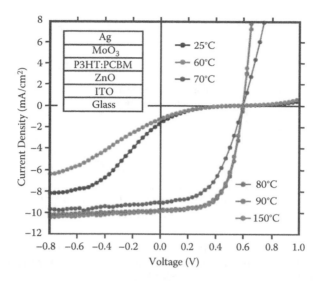

FIGURE 6.7 J–V curves of a P3HT:PCBM inverted solar cells with varying ZnO thicknesses. (Reproduced with permission from Y. Ka et al. 2012. *Organic Electronics* 14:100–104.)

the superior performance at higher temperatures is possibly caused by the alignment of energy levels. This implies that a large energy barrier exists for photogenerated electron extraction from LUMO of PCBM to the ZnO at lower temperatures [89].

Small et al. [90] incorporated a sol-gel film of ZnO–poly(vinyl pyrrolidone) (PVP) nanocomposite as electron selective layer. The ZnO–PVP nanocomposite synthetic route allows the size and concentration of the nanoclusters to be desirably controlled. In addition, the problem of nanoparticle aggregation in ZnO sol-gel without the PVP is avoided. They used a low-bandgap polymer polydithienogermole-thienopyrrolodione (PDTG-TPD) as the photoactive layer material. Initial light illumination resulted in J_{sc} = 10.9 mA/cm^2 and ff ~ 25.5%. However, with continuous light exposure, device performance increased significantly, eventually reaching J_{sc} = 12.9 mA/cm^2 and ff ~ 63.7%. The authors further subjected the surface of the ZnO-PVP nanocomposite with UV-ozone treatment for different times of 5, 10, 20, and 30 min in order to remove PVP from the surface. The 10 min UV-ozone resulted in optimal performance with J_{sc} = 14.4 mA/cm^2, V_{oc} = 0.86 V, ff = 68.8%, and PCE = 8.5% for the best cell. At 5 min, the ff reduced; this was attributed to incomplete removal of PVP from the surface of the ZnO film. Beyond 10 min, the ff also decreased due to excess oxygen adsorbed on the surface of the ZnO film. AFM topography investigation of the as-prepared and UV-ozone-treated surface resulted in root mean square (r.m.s.) roughness of 7.07 and 9.18 nm, respectively. The increase in r.m.s roughness with UV-ozone treatment further suggested the removal of PVP from the surface of the ZnO nanoclusters [90].

Adhikary et al. [91] studied the dependence of device performance with UV-ozone treatment on ZnO films prepared from sol-gel ZnO without PVP. The device structure of ITO/ZnO/(diketopyrrolopyrrole-terthiophene)(PDPP3T):PC60BM/MoO$_3$/Ag was used for this purpose. Figure 6.8 shows a J–V characteristic curve of PDPP3T:PC$_{60}$BM

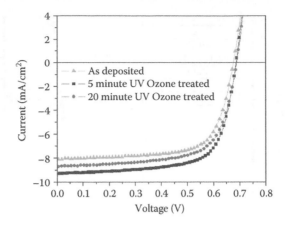

FIGURE 6.8 J–V plots of PDPP3T:PC60BM solar cells at different UV ozone treatment times on ZnO films. (Reproduced with permission from P. Adhikary et al. 2013. *IEEE Transactions on Electron Devices* 60(5):1763–1768.)

with different UV-ozone treatment time on ZnO films. ZnO films treated with UV-ozone yielded higher device performances than the as-deposited ZnO films. An efficiency of 4.45% with J_{sc} = 9.3 mA/cm^2, V_{oc} = 0.69 V, and ff = 69.5% was obtained for 5 min UV-ozone treated devices. Increased J_{sc} and fill factor for 5 and 20 min UV-ozone treated devices is supposed to be due to removal of organic solvents from the ZnO film. Longer UV-ozone treatment time (20 min) showed decrease in J_{sc} compared to 5 min, which contributed to the increased p-type defect (oxygen interstitials) formed in ZnO films. These defects act as a charge trapping site, reducing the electron extraction efficiency [91].

6.5.2   TiO$_2$

In 2009, Kim et al. [34] reported photovoltaic behavior of inverted solar cells under different times of light soaking employing solution-processed titania as electron transport layer (ETL). The device architecture is glass/ITO/TiO$_x$/P3HT:PCBM/Au. The J–V characteristics of devices are shown in Figure 6.9. The device performance as shown in Figure 6.9(a) resulted in J_{sc} = 1.57 mA/cm^2, V_{oc} = 0.36 V, ff = 0.21, and PCE = 0.09%. Under continuous illumination in air, the device performance improved and saturated after 10 min with J_{sc} = 8.02 mA/cm^2, V_{oc} = 0.38 V, ff = 0.51, and PCE = 1.5%. However, a reference cell without incorporation of the titania layer resulted in J_{sc} = 3.82 mA/cm^2, V_{oc} = 0.12 V, ff = 0.17, and PCE = 0.10%, as illustrated in Figure 6.9(b).

The photovoltaic behavior without titania did not exhibit any significant change, even with continuous illumination, suggesting that the titania layer must be responsible for photovoltaic behavior improvement. They explained that illumination excited electrons from the valence band into the conduction band in titania. The excited electrons, however, first filled the shallow electron traps in titania; continuous illumination resulted in total filling of these traps, causing saturation. The direct consequence of

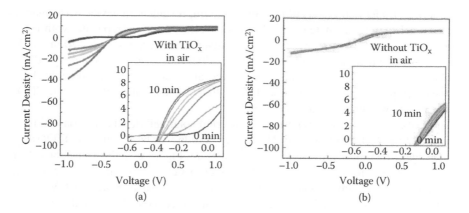

FIGURE 6.9 J–V curves of (a) titania incorporated and (b) titania absent inverted solar cells measured in air. (Reproduced with permission from K. Chang Su et al. 2009. *Applied Physics Letters* 94:113302.)

improved titania photoconductivity is the reduction in series resistance and increase in J_{sc} as observed in Figure 6.9(a) [34]. In 2011, Kang et al. [20] studied the ETL crystallinity effect on the J–V characteristics of organic solar cells. ALD deposited a 10 and 20 nm thick TiO_x layer as electron transport layer to provide conformal and pin-hole-free TiO_x film. The inverted device has a structure of glass/ITO/TiO_x/P3HT:PCBM/PEDOT:PSS/Ag. For the 10 nm TiO_x device, continuous illumination improved device performance (J_{sc} and V_{oc}), which eventually saturated after 20 min. However, there was no significant improvement for the 20 nm TiO_x device under prolonged illumination.

The absence of improved photovoltaic behavior in the 20 nm TiO_x film was studied through x-ray diffraction (XRD) analysis. Figure 6.10 shows the XRD pattern of the as-deposited 10 and 20 nm TiO_x film. No clear peak could be observed in the 10 nm film in Figure 6.10, supporting that the 10 nm TiO_x film had an amorphous structure, while the 20 nm film exhibited clear peaks, indicating suitable crystalline

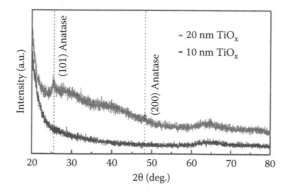

FIGURE 6.10 XRD spectra of the 10 and 20 nm TiO_x film. (Reproduced with permission from K. Yong-Jin et al. 2011. *Applied Physics Letters* 99:073308.)

properties. Under continuous illumination, the improvement in device performance of 10 nm TiO_x was attributed to the filling of shallow electron trap states within the amorphous phase TiO_x. However, such improvement was not seen for 20 nm TiO_x as it contained a crystalline phase. A critical thickness is necessary for crystalline behavior to emerge. This critical thickness was also believed to depend on the TiO_x growth temperature but in the work of Yong-Jin et al., the critical thickness for TiO_x crystallinity was greater than 10 nm [35].

In 2012, Kim et al. [88] reported the inverted solar cell dependence on UV irradiation of the TiO_x/ITO cathode. Their investigation was based on glass/ITO/TiO_x/ P3HT:PCBM/PEDOT:PSS/Ag device architecture. Figure 6.11 illustrates the effect of different light irradiation on the J–V characteristics. As shown in Figure 6.11(a), the device without any special light illumination results in an S-shaped J–V curve; however, white light irradiation for about 20 min recovered the normal J–V curve without any S-shape.

In order to understand the origin of white light soaking, Kim et al. filtered AM 1.5 G solar spectrum to expose the as-prepared devices to red, green, blue, and UV light. As depicted in Figure 6.11(b), the S-shaped behavior persisted after exposing the device to red, green, and blue light. However, the device showed significant improvement in J–V curves after UV irradiation. The performance improvement resulted from reduced series resistance of the ITO/TiO_x interface. Figure 6.12 shows the energy diagram of the ITO/TiO_x interface before and after UV irradiation. Exposure of TiO_x to air during device fabrication absorbs oxygen [$O_2(g)$], forming O_2^-. This O_2^- acts as trap sites affecting the charge transport properties. As shown in Figure 6.12(a), a wide Schottky barrier (E_B) is formed at the ITO/TiO_x interface because of energy difference between the work function of ITO (4.8 eV) and the work function of TiO_x (4.45 eV). Since photogenerated electrons cannot tunnel through this thick width of E_B, they are accumulated at the interface, forming an S-shaped J–V curve. However, after UV irradiation, electron–hole pairs are generated and the photogenerated holes combine with O_2^- (i.e., trap sites), leaving photogenerated electrons as free electrons; this increases the charge carrier density of TiO_x. This increase in

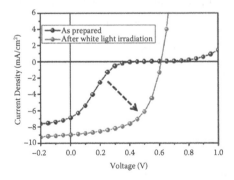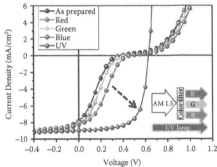

FIGURE 6.11 The influence of (a) white light, and (b) red, green, blue, and UV light on the J–V behavior of an inverted polymer solar cell incorporating TiO_x electron transport layer. (Reproduced with permission from J. Kim et al. 2012. *Journal of Applied Physics* 111:114511–114519.)

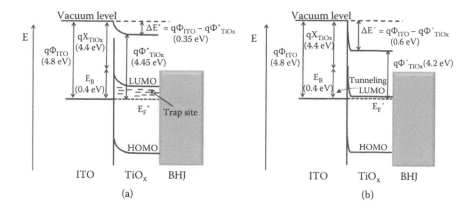

FIGURE 6.12 The energy diagram of ITO/TiO$_x$ interface (a) before UV irradiation, and (b) after UV irradiation. (Reproduced with permission from J. Kim et al. 2012. *Journal of Applied Physics* 111:114511–114519.)

charge carrier density induces the rearrangement of Fermi level at the interface of ITO and TiO$_x$, narrowing the E$_B$ width as shown in Figure 6.12(b). Reduction of E$_B$ width supports the accumulated electrons to tunnel through the barrier, enhancing the device performances [88].

6.5.3 Nb$_2$O$_5$

Solution-processed ZnO and TiO$_x$ thin films are normally used as an ETL for inverted polymer solar cells. These metal oxide layers are very transparent to the visible and near infrared regions as they have large bandgaps. However, the preparation of TiO$_x$ and ZnO is reported to be relatively complicated. TiO$_x$ sol-gel solution preparation requires a long time, several steps, and an inert atmosphere. Similarly, ZnO nanoparticle size is hard to control and needs moderate time for preparation. Siddiki, Venkatesan, and Qiao [93] reported Nb$_2$O$_5$ as a new ETL for double junction PSCs with a relatively simple process. At room temperature, niobium ethoxide (Nb(OC$_2$H$_5$)$_5$) precursor was mixed with ethanol and acetic acid to prepare the Nb$_2$O$_5$ sol-gel solution [92]. This precursor mixture was stirred for an hour and left for a day. Further, the solution was diluted with ethanol before spin coating. The spin-coated Nb$_2$O$_5$ sol-gel solution showed good attachability on the top of polymer:fullerene active layer [92,93].

Wiranwetchayan et al. [94] investigated the role of oxide thin dense layers Nb$_2$O$_5$ and TiO$_2$ in inverted structure PSCs. A power conversion of 2.7% and 2.8% was achieved for Nb$_2$O$_5$- and TiO$_2$-based inverted solar cells, respectively. As shown in Figure 6.13, the conduction band (CB) of TiO$_2$ is lower than the LOMO energy level of PCBM, allowing electron transport from PCBM to TiO$_2$. However, CB of Nb$_2$O$_5$ is higher than the LUMO of PCBM, which denotes an obstacle for electron transport from PCBM to Nb$_2$O$_5$. But the power conversion of 2.7% with Nb$_2$O$_5$ as buffer layer suggested that the electron transfer from PCBM to the collecting electrode should be through a tunneling process. They also showed that further increase in thickness

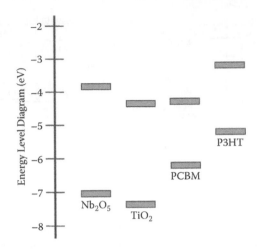

FIGURE 6.13   Energy-level diagram of Nb_2O_5, TiO_2, PCBM, and P3HT.

of Nb_2O_5 beyond tunneling distance degrades the device performance. For effective inverted structure PSCs, their experiment demonstrated that a thin oxide film is necessary to be deposited on FTO to prevent direct contact of FTO and P3HT. A thin oxide layer will enhance the formation of a continuously uniform PCBM film blocking the recombination of holes from P3HT in collecting electrodes. In conclusion, despite easy processing, Nb_2O_5 as an electron transport layer lags behind in terms of device performance [43].

6.5.4   NONCONJUGATED POLYELECTROLYTES AS A CATHODE INTERFACIAL LAYER

In order for PSCs to be viable for commercialization, long-term stability and large-scale printability requirements make an inverted device structure an advantageous approach. However, the high work function of the bottom ITO cathode that is ~4.8 eV forms a Schottky barrier with the LUMO level of fullerene. Inorganic metal oxides, such as titanium dioxide (TiO_2) or zinc oxide (ZnO), have been widely used as interfacial materials. These metal oxides require high annealing temperatures (over 200°C) to achieve crystallinity and sometimes require a post-UV treatment to enhance device performance [20,88,90,94]. For typical printing processes, high-temperature annealing of metal oxides is a limiting factor to flexible plastic substrates. Post-UV treatment of metal oxides in devices causes photo-oxidation of conjugated polymers, degrading device performance.

To overcome these issues of metal oxide layer, polymer-based interfacial materials, such as conjugated polyelectrolytes (CPEs), nonconjugated polyethylene oxide (PEO), and NPEs have been used. Kang et al. [20] demonstrated two NPEs, polyethyleneimine (PEI) and polyallylamine (PAA) containing nonconjugated backbones and functional amines. Polymer-based interlayers around 1–2 nm typically form ultrathin interfacial dipoles. These NPEs modified the work function of ITO from 4.8 to 4.2 eV and 4.0 eV for PAA and PEI, respectively. A Schottky barrier between ITO and $PC_{71}BM$ was successfully transformed to an ohmic contact in

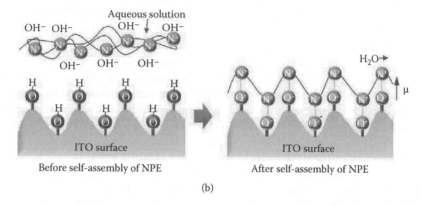

FIGURE 6.14 (a) Chemical structures of NPEs (PEI and PAA), and (b) before and after self-assembly of NPEs on ITO surface. (Reproduced with permission from H. Kang et al. 2012. *Advanced Materials* 24:3005–3009.)

order to achieve higher device performance. As a result of the Schottky-to-ohmic transformation, device performance became UV independent with high PCEs of 6.3%. ITO/NPEs also showed similar optical and morphological properties as those of bare ITO, which is a promising aspect for application in inverted PSCs. Figure 6.14(a) shows the chemical structures of PEI and PAA. Functional amines on NPEs offer two advantages: (1) they are water soluble, enabling solution processing and multilayer stacking without intermixing with BHJ composites; and (2) they can be protonated in aqueous solution by accepting protons (H⁺) dissociated from water. Figure 6.14(b) illustrates that NPEs create electrostatic self-assembly and form strong dipoles across the interface between NPE and ITO surfaces. These surface dipole moments pointing outward from ITO can reduce the work function of ITO through a downward vacuum level shift [19,20].

6.6 DIFFERENT ANODE INTERFACIAL LAYERS (HOLE TRANSPORT LAYERS) USED IN INVERTED SOLAR CELLS

A buffer layer is generally placed between the active layer and high work function metal anode. Such a layer multifunctions—for example, as hole transport layer, electron block layer, exciton block layer, and protection layer for the underlying active layer. Some of these layers include MoO_3, V_2O_5, WO_3, PEDOT:PSS, and CuO_x. These buffer layers should be properly matched with the energy levels of polymers and anode materials and possess good hole transporting capability. In addition, the thickness of these layers should be optimized as it contributes to the device series resistance. Higher thickness may result in lower *ff* and eventually decrease the device performance [31,40,41,95]. PEDOT:PSS is a widely used hole transport layer in PSCs [96]. However, for inverted PSCs, MoO_3 as hole transport layer has shown tremendous progress recently.

Zhao et al. [97] demonstrated that the MoO_3 layer between the active layer and Ag anode improves both fill factor and photocurrent. It also works as an optical spacer to adjust optical field distribution. A PCE of 3.86% was achieved with $P3HT:PC_{61}BM$-based active layer [97]. Sun et al. [73] reported a PCE of inverted solar cell over 4% using solution-processed MoO_x as anodic modification, which is comparable to their thermally deposited MoO_x. Using air plasma treatment, MoO_3 was reduced to MoO_x, where x < 3 and got similar device performance as with thermally deposited MoO_x [73]. Liu et al. [98] showed a new concept utilizing a work-function tunable transparent MoO_3–Al composite layer as cathode buffer layer and MoO_3/Al as the anode. Inverted PSCs based on $PCDTBT:PC_{70}BM$ blend showed a high PCE of 6.77% with an *ff* of 70.7%. Because this device structure involves a simple combination of only two materials (MoO_3 and Al) as both the cathode and anode buffer layer, it may ease device fabrication and reduce cost in large-scale application [98]. Although the vacuum evaporation of hole transport layers (HTLs) (e.g., MoO_3 and V_2O_5) gives higher performance, development in solution processing of these layers has great importance as it can eliminate the need for a high vacuum evaporation process, which can ultimately reduce the overall cost and even ease the steps for roll-to-roll processing.

Lin et al. [40] showed that sol-gel derived copper oxide (CuO_x) as an interlayer at the organic/metal interface. This interlayer can form an ohmic contact between the polymer and electrode, improving the *ff*, R_{sh}, and R_s of the device. Furthermore, CuO_x can also enhance light absorption by redistributing light intensity [40]. Vanadium pentoxide (V_2O_5) can also be used as hole transport layer. The Yang group demonstrated inverted and regular PSCs with V_2O_5 as hole transport layer. In regular device structure, they even compared the performance of PEDOT:PSS, V_2O_5, and MoO_3 as hole selective layers. Devices with MoO_3 as a hole selective layer showed a higher PCE of 3.33% compared to 3.18% and 3.10% of PEDOT:PSS- and V_2O_5-based devices, respectively [41]. In another study, they demonstrated an inverted device structure with V_2O_5 as hole selective layer with a PCE of 2.25% [31]. Yusoff, Kim, and Jang [99] showed spin-coated V_2O_5 and ZnO nanoparticles as hole and electron extraction layers. A PCE of 5.53% was obtained for the P3HT:ICBA-based regular device structure [99]. Kettle et al. [39] studied three hole transport layers PEDOT:PSS, sputtered nickel oxide (NiO), and MoO_3 in PCPDTBT:C71-PCBM-based regular solar cells. In their experiments, PCEs of 3.89%, 3.74%, and 3.84% were demonstrated for sputtered NiO-, PEDOT:PSS-, and MoO_3-based devices, respectively.

Hancox et al. [74] reported a solution-processable V_2O_X hole extraction layer. Devices based on a P3HT:PCBM active layer generated a PCE of 3.17%, 3.34%, and 3.31% for hole extraction layers of PEDOT:PSS, V_2O_X (solution processed), and MoO_x respectively. Under 60 min of continuous illumination at 100 mW/cm^2 AM 1.5G in N$_2$ atmosphere, devices with V_2O_x, MoO_x, and PEDOT:PSS retained 89%, 94%, and 68% of the starting efficiencies, respectively [74]. Espinosa et al. [44] demonstrated hydrated V_2O_5 as a hole transport layer for full roll-to-roll processing. A device structure of PET/ITO/ZnO/P3HT:PCBM/V_2O_5.(H$_2$O)n/Ag was employed. ITO and Ag were deposited using a screen printing technique while ZnO, P3HT:PCBM and V_2O_5 were deposited using slot-die coating. Modules containing 16 serially connected cells with an active device area of 360 cm^2 yielded a PCE up to 0.18% for the roll-to-roll processing technique only [44]. Tungsten oxide (WO$_3$) has also been applied as an anode interfacial layer between the photoactive layer and top electrode in inverted PSCs.

By applying 10 nm of WO$_3$ between the active layer and Ag anode, Tao et al. [42] demonstrated a PCE of 2.58% for inverted PSC with an ITO/nc-TiO$_2$/P3HT:PCBM/WO$_3$/Ag device structure. WO$_3$ has a high work function of −4.8 eV making WO$_3$ capable of collecting hole and blocking electrons. In addition, they also demonstrated a transparent inverted PSC with ITO/nc-TiO$_2$/P3HT:PCBM/WO$_3$(10 nm)/Ag(13 nm)/WO$_3$(40 nm) where the 10 nm WO$_3$ layer acts as a hole transport layer and both ITO and Ag/WO$_3$ act as transparent electrodes. When illuminated from the ITO side, a PCE of 1.80% was achieved. Similarly, a PCE of 0.96% was achieved when illuminated from the Ag/WO$_3$ side. This concept of transparent inverted PSC structure can be utilized for the design of tandem solar cells where multiple photoactive layers can be employed to absorb a wide range of the solar spectrum [42]. A solution-processed WO$_3$ (s-WO$_3$) as anode buffer layer has also been demonstrated by Tan et al. [75]. This s-WO$_3$ showed higher transparency in the visible range and higher hole mobility of 9.4×10^{-3}cm^2/V·s, which is advantageous for effective hole collection and transportation. Applying this s-WO$_3$ anode buffer layer in a regular structure based on P3HT/IC$_{70}$BA, a PCE of 6.36% was demonstrated under the illumination of AM 1.5G, 100 mW/cm^2 [75].

6.7 CONCLUSION

The inverted bulk heterojunction solar cell is emerging as a strong alternative to regular solar cells. Due to advantages of inherent vertical phase segregation and the capability of giving higher efficiency and stability, recently more research has been based on inverted solar cells. The interfacial layer has a great impact on the performance and stability of the inverted solar cells. Lots of materials for interfacial layers have been studied for better aligning the work function with the active layer and electrodes, efficient charge transfer, surface and work function modification, enhancement of optical properties, and improving the quality and easiness of processing. Some widely used materials and their effects have been discussed in this chapter. Whatever the device configuration is, some key factors that need to be focused on are synthesis of low-bandgap polymers to increase J_{sc}, appropriate lowering of HOMO level of donor material to realize higher V_{oc}, search for and selection of materials to align work functions of different layers effectively, and design of tandem structures to harvest a wide range of solar spectra. In addition, more research has to be done to further improve the stability and processing to favor mass production and cost reduction. Integrating these factors, organic solar cells can move closer to realize the target efficiency (15%) and stability to compete with the existing technology.

ACKNOWLEDGMENT

This work was supported in part by NSF CAREER (ECCS-0950731), NSF EPSCoR (Grant No. 0903804) and the state of South Dakota, NASA EPSCoR (No. NNX13AD31A), 3M nontenured faculty award, and SDBoR CRGP grant.

REFERENCES

1. M. Siddiki, J. Li, D. Galipeau, and Q. Qiao. 2010. A review on polymer multijunction solar cells. *Energy and Environmental Sciences* 3:867–883.
2. M. K. Siddiki, S. Venkatesan, D. Galipeau, and Q. Qiao. 2013. Kelvin probe force microscopic imaging of the energy barrier and energetically favorable offset of interfaces in double-junction organic solar cells. *ACS Applied Materials & Interfaces* 5:1279–1286.
3. M. K. Siddiki, S. Venkatesan, M. Wang, and Q. Qiao. 2013. Materials and devices design for efficient double junction polymer solar cells. *Solar Energy Materials and Solar Cells* 108:225–229.
4. P. Taranekar, Q. Qiao, J. Jiang, K. S. Schanze, and J. R. Reynolds. 2007. Hyperbranched conjugated polyelectrolyte bilayers for solar cell applications. *Journal of the American Chemical Society* 129:8958–8959.
5. Y. Xie, Y. Bao, J. Du, C. Jiang, and Q. Qiao. 2012. Understanding of morphology evolution in local aggregates and neighboring regions for organic photovoltaics. *Physical Chemistry and Chemical Physics* 14:10168–10177.
6. Y. Xie, Y. Li, L. Xiao, Q. Qiao, R. Dhakal, Z. Zhang, et al. 2010. Femtosecond time-resolved fluorescence study of P3HT/PCBM blend films. *Journal of Physical Chemistry C* 114:14590–14600.
7. T. Xu, Q. Chen, D.-H. Lin, H.-Y. Wu, C.-F. Lin, and Q. Qiao 2011. Self-assembled thienylsilane molecule as interfacial layer for ZnO nanowire/polymer hybrid system. *Journal of Photonics for Energy* 1:011107.
8. T. Xu and Q. Qiao. 2011. Conjugated polymer-inorganic semiconductor hybrid solar cells. *Energy & Environmental Science* 4:2700–2720.
9. T. Xu, S. Venkatesan, D. Galipeau, and Q. Qiao. 2011. Study of polymer/ZnO nanostructure interfaces by kelvin probe force microscopy. *Solar Energy Materials and Solar Cells* 108:246–251.
10. T. Xu, M. Yan, J. Hoefelmeyer, and Q. Qiao. 2011. Exciton migration and charge transfer in chemically linked P3HT-TiO2 nanorod composite. *RSC Advance* 2:854–862.
11. S. Shao, K. Zheng, T. Pullerits, and F. Zhang. 2012. Enhanced performance of inverted polymer solar cells by using poly(ethylene oxide)-modified ZnO as an electron transport layer. *ACS Applied Materials & Interfaces* 5:380–385.
12. J. Li, M. Yan, Y. Xie, and Q. Qiao. 2011. Linker effects on optoelectronic properties of alternate donor–acceptor conjugated polymers. *Energy Environmental Science* 4:4276–4283.
13. S. R. Forrest. 2004. The path to ubiquitous and low-cost organic electronic appliances on plastic. *Nature* 428:911–918.
14. K. M. Vaeth. 2003. OLED-display technology-OLEDs are exciting not only because of their excellent front-of-screen performance, but because their flexibility invites many approaches to materials and device design. *Information Display* 19:12–17.

15. D. J. Gundlach, Y. Y. Lin, T. N. Jackson, S. F. Nelson, and D. G. Schlom. 1997. Pentacene organic thin-film transistors-molecular ordering and mobility. *IEEE Electron Device Letters* 18:87–89.

16. G. Li, R. Zhu, and Y. Yang. 2012. Polymer solar cells. *Nature Photonics* 6:153–161.

17. Y. Liang, Z. Xu, J. Xia, S.-T. Tsai, Y. Wu, G. Li, et al. 2010. For the bright future—Bulk heterojunction polymer solar cells with power conversion efficiency of 7.4%. *Advanced Materials* 22:E135–E138.

18. J. Peet, M. L. Senatore, A. J. Heeger, and G. C. Bazan. 2009. The role of processing in the fabrication and optimization of plastic solar cells. *Advanced Materials* 21:1521–1527.

19. Y. Zhou, C. Fuentes-Hernandez, J. Shim, J. Meyer, A. J. Giordano, H. Li, et al. 2012. A universal method to produce low–work function electrodes for organic electronics. *Science* 336:327–332.

20. H. Kang, S. Hong, J. Lee, and K. Lee. 2012. Electrostatically self-assembled nonconjugated polyelectrolytes as an ideal interfacial layer for inverted polymer solar cells. *Advanced Materials* 24:3005–3009.

21. J. Y. Kim, S. H. Kim, H. H. Lee, K. Lee, W. Ma, X. Gong, et al. 2006. New architecture for high-efficiency polymer photovoltaic cells using solution-based titanium oxide as an optical spacer. *Advanced Materials* 18:572–576.

22. G. Li, C. W. Chu, V. Shrotriya, J. Huang, and Y. Yang. 2006. Efficient inverted polymer solar cells. *Applied Physics Letters* 88:253503-3.

23. F. J. Zhang, D. W. Zhao, Z. L. Zhuo, H. Wang, Z. Xu, and Y. S. Wang. 2010. Inverted small molecule organic solar cells with Ca modified ITO as cathode and MoO3 modified Ag as anode. *Solar Energy Materials and Solar Cells* 94:2416–2421.

24. F. Zhang, X. Xu, W. Tang, J. Zhang, Z. Zhuo, J. Wang, et al. 2011. Recent development of the inverted configuration organic solar cells. *Solar Energy Materials and Solar Cells* 95:1785–1799.

25. X. Gong. 2012. Towards high performance inverted polymer solar cells. *Polymer* 53:5437–5448.

26. S. K. Hau, H.-L. Yip, and A. K. Y. Jen. 2010. A review on the development of the inverted polymer solar cell architecture. *Polymer Reviews* 50:474–510.

27. Z. Xu, Chen, L.-M., Yang, G., Huang, C.-H., Hou, J., Wu, Y., Li, G., Hsu, C.-S. and Yang, Y. 2009. Vertical phase separation in poly(3-hexylthiophene): Fullerene derivative blends and its advantage for inverted structure solar cells. *Advances in Functional Materials* 19:1227–1234.

28. M. De Jong, L. Van Ijzendoorn, and M. De Voigt. 2000. Stability of the interface between indium-tin-oxide and poly (3, 4-ethylenedioxythiophene)/poly (styrenesulfonate) in polymer light-emitting diodes. *Applied Physics Letters* 77:2255–2257.

29. S. K. Hau, H.-L. Yip, N. S. Baek, J. Zou, K. O'Malley, and A. K. Y. Jen. 2008. Air-stable inverted flexible polymer solar cells using zinc oxide nanoparticles as an electron selective layer. *Applied Physics Letters* 92:253301–253303.

30. C. Waldauf, M. Morana, P. Denk, P. Schilinsky, K. Coakley, S. Choulis, et al. 2006. Highly efficient inverted organic photovoltaics using solution based titanium oxide as electron selective contact. *Applied Physics Letters* 89:233517.

31. G. Li, C. W. Chu, V. Shrotriya, J. Huang, and Y. Yang. 2006. Efficient inverted polymer solar cells. *Applied Physics Letters* 88:253503.

32. Y. Sun, J. H. Seo, C. J. Takacs, J. Seifter, and A. J. Heeger. 2011. Inverted polymer solar cells integrated with a low-temperature-annealed sol-gel-derived ZnO film as an electron transport layer. *Advanced Materials* 23:1679–1683.

33. A. K. K. Kyaw, X. W. Sun, C. Y. Jiang, G. Q. Lo, D. W. Zhao, and D. L. 2008. An inverted organic solar cell employing a sol-gel derived ZnO electron selective layer and thermal evaporated [MoOsub 3] hole selective layer. *Applied Physics Letters* 93:221107.

34. K. Chang Su, S. S. Lee, E. D. Gomez, K. Jong Bok, and L. Yueh-Lin. 2009. Transient photovoltaic behavior of air-stable, inverted organic solar cells with solution-processed electron transport layer. *Applied Physics Letters* 94:113302.

35. K. Yong-Jin, K. Chang Su, Y. Dae Sung, J. Sung Hoon, L. Kyounga, K. Do-Geun, et al. 2011. Effect of electron transport layer crystallinity on the transient characteristics of inverted organic solar cells. *Applied Physics Letters* 99:073308.

36. H. Cheun, C. Fuentes-Hernandez, Y. Zhou, W. J. Potscavage, S.-J. Kim, J. Shim, et al. 2010. Electrical and optical properties of ZnO processed by atomic layer deposition in inverted polymer solar cells. *Journal of Physical Chemistry C* 114:20713–20718.

37. L. Hua-Hsien, C. Li-Min, X. Zheng, L. Gang, and Y. Yang. 2008. Highly efficient inverted polymer solar cell by low temperature annealing of Cs2CO3 interlayer. *Applied Physics Letters* 92:173303.

38. T. Stubhan, H. Oh, L. Pinna, J. Krantz, I. Litzov, and C. J. Brabec. 2011. Inverted organic solar cells using a solution processed aluminum-doped zinc oxide buffer layer. *Organic Electronics* 12:1539–1543.

39. J. Kettle, H. Waters, M. Horie, and S. Chang. 2012. Effect of hole transporting layers on the performance of PCPDTBT: PCBM organic solar cells. *Journal of Physics D: Applied Physics* 45:125102.

40. M.-Y. Lin, C.-Y. Lee, S.-C. Shiu, I.-J. Wang, J.-Y. Sun, W.-H. Wu, et al. 2010. Sol–gel processed CuOx thin film as an anode interlayer for inverted polymer solar cells. *Organic Electronics* 11:1828–1834.

41. V. Shrotriya, L. Gang, Y. Yan, C. Chih-Wei, and Y. Yang. 2006. Transition metal oxides as the buffer layer for polymer photovoltaic cells. *Applied Physics Letters* 88:073508.

42. C. Tao, S. Ruan, G. Xie, X. Kong, L. Shen, F. Meng, et al. 2009. Role of tungsten oxide in inverted polymer solar cells. *Applied Physics Letters* 94:043311.

43. O. Wiranwetchanyan, Z. Liang, Q. Zhang, G. Cao, and P. Singjai. 2011. The role of oxide thin layer in inverted structure polymer solar cells. *Materials Sciences and Applications* 2:1697–1701

44. N. Espinosa, H. F. Dam, D. M. Tanenbaum, J. W. Andreasen, M. Jørgensen, and F. C. Krebs. 2011. Roll-to-Roll processing of inverted polymer solar cells using hydrated vanadium (V) oxide as a PEDOT:PSS replacement. *Materials* 4:169–182.

45. F. C. Krebs, S. A. Gevorgyan, and J. Alstrup. 2009. A roll-to-roll process to flexible polymer solar cells: Model studies, manufacture and operational stability studies. *Journal of Materials Chemistry* 19:5442–5451.

46. C. Deibel, and V. Dyakonov. 2010. Polymer–fullerene bulk heterojunction solar cells. *Reports on Progress in Physics* 73:096401.

47. X. Y. Zhu, Q. Yang, and M. Muntwiler. 2009. Charge-transfer excitons at organic semiconductor surfaces and interfaces. *Accounts of Chemical Research* 42:1779–1787.

48. P. Heremans, D. Cheyns, and B. P. Rand. 2009. Strategies for increasing the efficiency of heterojunction organic solar cells: Material selection and device architecture. *Accounts of Chemical Research* 42:1740–1747.

49. W. Brutting, ed. 2007. *Physics of organic semiconductors*. Weinheim, Germany: Wiley-VCH.

50. E. Bundgaard, and F. C. Krebs. 2007. Low band gap polymers for organic photovoltaics. *Solar Energy Materials and Solar Cells* 91:954–985.

51. L. Dou, J. You, J. Yang, C.-C. Chen, Y. He, S. Murase, et al. 2012. Tandem polymer solar cells featuring a spectrally matched low-bandgap polymer. *Nature Photonics* 6:180–185.

52. D. E. Markov, E. Amsterdam, P. W. M. Blom, A. B. Sieval, and J. C. Hummelen 2005. Accurate measurement of the exciton diffusion length in a conjugated polymer using a heterostructure with a side-chain cross-linked fullerene layer. *Journal of Physical Chemistry A* 109:5266–5274.

53. S. Gunes, H. Neugebauer, and N. S. Sariciftci. 2007. Conjugated polymer-based organic solar cells. *Chemical Reviews* 107:1324–1338.

54. P. E. Shaw, A. Ruseckas, and I. D. W. Samuel. 2008. Exciton diffusion measurements in poly(3-hexylthiophene). *Advanced Materials* 20:3516–3520.
55. B. R. Saunders, and M. L. Turner. 2008. Nanoparticle–polymer photovoltaic cells. *Advances in Colloid and Interface Science* 138:1–23.
56. N. S. Sariciftci, L. Smilowitz, A. J. Heeger, and F. Wudl. 1992. Photoinduced electron transfer from a conducting polymer to Buckminster fullerene. *Science* 258:1474–1476.
57. C. J. Brabec, N. S. Sariciftci, and J. C. Hummelen. 2001. Plastic solar cells. *Advanced Functional Materials* 11:15–26.
58. M. M. Mandoc, L. J. A. Koster, and P. W. M. Blom. 2007. Optimum charge carrier mobility in organic solar cells. *Applied Physics Letters* 90:133504 .
59. R. N. Marks, J. J. M. Halls, D. D. C. Bradley, R. H. Friend, and A. B. Holmes. 1994. The photovoltaic response in poly(p-phenylene vinylene) thin-film devices. *Journal of Physics-Condensed Matter* 6:1379–1394.
60. C. W. Tang. 1986. Two-layer organic photovoltaic cell. *Applied Physics Letters* 48:183–185.
61. H. Hoppe, and N. Sariciftci. Polymer solar cells. *Photoresponsive Polymers II* 214:1–86.
62. M. Hiramoto, H. Fujiwara, and M. Yokoyama. 1991. Three-layered organic solar cell with a photoactive interlayer of codeposited pigments. *Applied Physics Letters* 58:1062–1064.
63. M. Hiramoto, H. Fujiwara, and M. Yokoyama. 1992. p-i-n like behavior in three-layered organic solar cells having a co-deposited interlayer of pigments. *Journal of Applied Physics* 72:3781–3787.
64. G. Yu, J. Gao, J. C. Hummelen, F. Wudl, and A. J. Heeger. 1995. Polymer photovoltaic cells—Enhanced efficiencies via a network of internal donor–acceptor heterojunctions. *Science* 270:1789–1791.
65. Y. Wang, X. Chen, Y. Zhong, F. Zhu, and K. P. Loh. 2009. Large area, continuous, few-layered graphene as anodes in organic photovoltaic devices. *Applied Physics Letters* 95:063302.
66. J. Van De Lagemaat, T. M. Barnes, G. Rumbles, S. E. Shaheen, T. J. Coutts, C. Weeks, et al. 2006. Organic solar cells with carbon nanotubes replacing InO: Sn as the transparent electrode. *Applied Physics Letters* 88:233503.
67. X. Wang, L. Zhi, N. Tsao, Ž. Tomović, J. Li, and K. Müllen. 2008. Transparent carbon films as electrodes in organic solar cells. *Angewandte Chemie* 120:3032–3034.
68. J. Lee, D. Lee, D. Lim, and K. Yang. 2007. Structural, electrical and optical properties of ZnO: Al films deposited on flexible organic substrates for solar cell applications. *Thin Solid Films* 515:6094–6098.
69. Y. Kim, S. A. Choulis, J. Nelson, D. D. C. Bradley, S. Cook, and J. R. Durrant. 2005. Device annealing effect in organic solar cells with blends of regioregular poly(3-hexyl-thiophene) and soluble fullerene. *Applied Physics Letters* 86:063502.
70. M. Al-Ibrahim, O. Ambacher, S. Sensfuss, and G. Gobsch. 2005. Effects of solvent and annealing on the improve.ed performance of solar cells based on poly (3-hexylthio-phene): Fullerene. *Applied Physics Letters* 86:201120.
71. M. T. Rispens, A. Meetsma, R. Rittberger, C. J. Brabec, N. S. Sariciftci, and J. C. Hummelen. 2003. Influence of the solvent on the crystal structure of PCBM and the efficiency of MDMO-PPV:PCBM "plastic" solar cells. *Chemical Communications* 2003:2116–2118.
72. H. Hoppe, and N. S. Sariciftci. 2004. Organic solar cells: An overview. *Journal of Materials Research* 19:1924–1945.
73. J.-Y. Sun, W.-H. Tseng, S. Lan, S.-H. Lin, P.-C. Yang, C.-I. Wu, et al. 2013. Performance enhancement in inverted polymer photovoltaics with solution-processed MoO_x and air-plasma treatment for anode modification. *Solar Energy Materials and Solar Cells* 109:178–184.

74. I. Hancox, L. Rochford, D. Clare, M. Walker, J. Mudd, P. Sullivan, et al. 2012. Optimization of a high work function solution processed vanadium oxide hole-extracting layer for small molecule and polymer organic photovoltaic cells. *Journal of Physical Chemistry C* 117:49–57.

75. Z. A. Tan, L. Li, C. Cui, Y. Ding, Q. Xu, S. Li, et al. 2012. Solution-processed tungsten oxide as an effective anode buffer layer for high-performance polymer solar cells. *Journal of Physical Chemistry C* 116:18626–18632.

76. M. Glatthaar, M. Niggemann, B. Zimmermann, P. Lewer, M. Riede, A. Hinsch, et al. 2005. Organic solar cells using inverted layer sequence. *Thin Solid Films* 491:298–300.

77. Z. He, C. Zhong, S. Su, M. Xu, H. Wu, and Y. Cao. 2012. Enhanced power-conversion efficiency in polymer solar cells using an inverted device structure. *Nature Photonics* 6:591–595.

78. S. Y. Heriot, and R. A. Jones. 2005. An interfacial instability in a transient wetting layer leads to lateral phase separation in thin spin-cast polymer-blend films. *Nature Materials* 4:782–786.

79. C. M. Björström, A. Bernasik, J. Rysz, A. Budkowski, S. Nilsson, M. Svensson, et al. 2005. Multilayer formation in spin-coated thin films of low-bandgap polyfluorene: PCBM blends. *Journal of Physics: Condensed Matter* 17:L529.

80. J. Subbiah, C. M. Amb, J. R. Reynolds, and F. So. 2012. Effect of vertical morphology on the performance of silole-containing low-bandgap inverted polymer solar cells. *Solar Energy Materials and Solar Cells* 97:97–101.

81. Y.-I. Lee, J.-H. Youn, M.-S. Ryu, J. Kim, H.-T. Moon, and J. Jang. 2011. Highly efficient inverted poly (3-hexylthiophene): Methano-fullerene 6, 6]-phenyl C71-butyric acid methyl ester bulk heterojunction solar cell with Cs_2CO_3 and MoO_3. *Organic Electronics* 12:353–357.

82. T.-Y. Chu, S.-W. Tsang, J. Zhou, P. G. Verly, J. Lu, S. Beaupré, et al. 2012. High-efficiency inverted solar cells based on a low bandgap polymer with excellent air stability. *Solar Energy Materials and Solar Cells* 96:155–159.

83. K. H. Steven, Y. Hin-Lap, B. Nam Seob, Z. Jingyu, O. M. Kevin, and K. Y. J. Alex. 2008. Air-stable inverted flexible polymer solar cells using zinc oxide nanoparticles as an electron selective layer. *Applied Physics Letters* 92:253301.

84. M. S. White, D. C. Olson, S. E. Shaheen, N. Kopidakis, and D. S. Ginley. 2006. Inverted bulk-heterojunction organic photovoltaic device using a solution-derived ZnO underlayer. *Applied Physics Letters* 89:143517.

85. S. Cros, M. Firon, S. Lenfant, P. Trouslard, and L. Beck. 2006. Study of thin calcium electrode degradation by ion beam analysis. *Nuclear Instruments and Methods in Physics Research Section B: Beam Interactions with Materials and Atoms* 251:257–260.

86. S. K. Hau, H.-L. Yip, N. S. Baek, J. Zou, K. O'Malley, and A. K.-Y. Jen. 2008. Air-stable inverted flexible polymer solar cells using zinc oxide nanoparticles as an electron selective layer. *Applied Physics Letters* 92:253301.

87. F. Verbakel, S. C. J. Meskers, and R. A. J. Janssen. 2006. Electronic memory effects in diodes from a zinc oxide nanoparticle-polystyrene hybrid material. *Applied Physics Letters* 89:102103.

88. J. Kim, G. Kim, Y. Choi, J. Lee, S. H. Park, and K. Lee. 2012. Light-soaking issue in polymer solar cells: Photoinduced energy level alignment at the sol-gel processed metal oxide and indium tin oxide interface. *Journal of Applied Physics* 111:114511–114519.

89. Y. Ka, E. Lee, S. Y. Park, J. Seo, D.-G. Kwon, H. H. Lee, et al. 2012. Effects of annealing temperature of aqueous solution-processed ZnO electron-selective layers on inverted polymer solar cells. *Organic Electronics* 14:100–104.

90. C. E. Small, S. Chen, J. Subbiah, C. M. Amb, S.-W. Tsang, T.-H. Lai, et al. 2012. High-efficiency inverted dithienogermole-thienopyrrolodione-based polymer solar cells. *Nature Photonics* 6:115–120.

91. P. Adhikary, S. Venkatesan, P. P. Maharjan, D. Galipeau, and Q. Qiao. 2013. Enhanced performance of PDPP3T/PC$_{60}$BM solar cells using high boiling solvent and UV–ozone treatment. *IEEE Transactions on Electron Devices* 60 (5): 1763–1768.

92. M. Lira-Cantu, K. Norrman, J. W. Andreasen, and F. C. Krebs. 2006. Oxygen release and exchange in niobium oxide MEHPPV hybrid solar cells. *Chemistry of Materials* 18:5684–5690.

93. M. K. Siddiki, S. Venkatesan, and Q. Qiao. 2012. Nb2O5 as a new electron transport layer for double junction polymer solar cells. *Physical Chemistry Chemical Physics* 14:4682–4686.

94. O. Wiranwetchayan, Z. Liang, Q. Zhang, G. Cao, and P. Singjai. 2011. The role of oxide thin layer in inverted structure polymer solar cells. *Materials Sciences and Applications* 2:1697–1701.

95. F. Cheng, G. Fang, X. Fan, N. Liu, N. Sun, P. Qin, et al. 2011. Enhancing the short-circuit current and efficiency of organic solar cells using MoO$_3$ and CuPc as buffer layers. *Solar Energy Materials and Solar Cells* 95:2914–2919.

96. G. Li, V. Shrotriya, J. Huang, Y. Yao, T. Moriarty, K. Emery, et al. 2005. High-efficiency solution processable polymer photovoltaic cells by self-organization of polymer blends. *Nature Materials* 4:864–868.

97. D. Zhao, S. Tan, L. Ke, P. Liu, A. Kyaw, X. Sun, et al. 2010. Optimization of an inverted organic solar cell. *Solar Energy Materials and Solar Cells* 94:985–991.

98. J. Liu, S. Shao, G. Fang, B. Meng, Z. Xie, and L. Wang. 2012. High-efficiency inverted polymer solar cells with transparent and work-function tunable MoO3-Al composite film as cathode buffer layer. *Advanced Materials* 24:2774–2779.

99. A. R. B. Mohd Yusoff, H. P. Kim, and J. Jang. 2013. Organic photovoltaics with V2O5 anode and ZnO nanoparticles cathode buffer layers. *Organic Electronics* 14:858–861.

7 Monocrystalline Silicon Solar Cell Optimization and Modeling

Joanne Huang and Victor Moroz

CONTENTS

Abstract: The performance of a monocrystalline silicon solar cell is determined by a variety of optical and electronic physical mechanisms. Optimization of cell performance requires finding trade-offs for the competing physical mechanisms. In this chapter we use three-dimensional (3D) optical analysis and electronic transport analysis to determine optimization space for a solar cell with textured surface and point rear contacts. Optimization of the surface texture, antireflective layers, contact size and pitch, and junctions can improve the overall solar cell efficiency by over 4%. Comparison between three-dimensional analysis and the simplified two-dimensional and one-dimensional approaches is provided with recommended validity ranges for each approach.

7.1   INTRODUCTION

A solar cell is designed to trap as much sunlight as possible and to convert most of it into electricity. Light trapping is important in the wavelength range from 0.3 to 1.2 μm, which is where most of the solar irradiation energy is contained and which can be converted into optically generated free electrons and holes in silicon. The amount of light that can be trapped inside the solar cell is determined by the surface texture and by the contacts and antireflective layers that cover its front and rear surfaces. Usually, the smaller the contacts are, the better is light trapping, because front surface contacts act as a shadow and rear surface contacts degrade reflection of the long wavelengths.

Electrical efficiency of the solar cell also depends on the location and size of the front and rear contacts. Usually, the larger the contacts are, the better is conduction of the cell. This brings the demands of optical performance and the electrical performance of the cell into conflict. Modeling such competing physical mechanisms enables achieving a trade-off and optimization of the overall performance of the cell.

In this work, we demonstrate how three-dimensional simulation can be used to find trade-offs in combined optical and electrical performances of the solar cell. In addition to maximizing performance, we also address tightening of the performance spread to improve parametric yield.

The next section is dedicated to modeling optical effects, followed by a section that adds electronic effects and looks at the overall cell performance.

7.2   MODELING OPTICAL EFFECTS

7.2.1   Textured Surface

Most of the monocrystalline solar cells on the market have surface texture [1,2]. Typical measured texture consists of overlapping pyramids with random heights and locations.

The sizes of the pyramids vary from 2 to 20 μm depending on the etching process. All facets of the pyramids of any size have the same crystal orientation of {111}, which is the densest surface in silicon crystal lattice. The {111} facets have the same angle of 55° with respect to the wafer plane.

Area of the textured surface is $\sqrt{3} \approx 1.73$ times larger than area of a flat surface on which the pyramids are formed. This ratio is independent of the pyramid size or whether the pyramids are random or regular.

Typically, both the front and the rear surfaces of the silicon wafer are randomly textured, but here we will look at different cases, with flat, regularly textured, and randomly textured pyramids. An example of a randomly textured silicon surface used in simulations is shown in Figure 7.1.

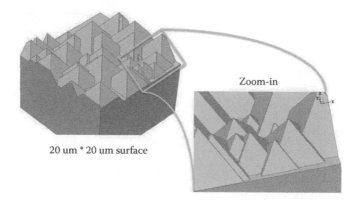

FIGURE 7.1 Simulated solar cell surface with random texture. Average pyramid height is 4.2 mm.

Typical pyramid height is considerably larger than the longest relevant light wavelength. However, different wavelengths in the solar spectrum exhibit very distinct optical behaviors in terms of refraction at the interfaces and in terms of how quickly they get absorbed in silicon. In this work, we use the Sentaurus tool suite [3] for optical and electrical analysis of the solar cells.

7.2.2 OPTICAL PERFORMANCE OF REGULAR SURFACE PATTERNS

Reflectance changes across the solar spectrum and is very sensitive to the surface texture and films deposited on the surface. Transmittance becomes nonzero only for the long wavelengths, because short waves do not get deep enough to reach the rear surface and escape from it.

Figure 7.2 illustrates reflectance as a function of the light wavelength for different wafer surfaces. The legend describes the top surface of the wafer before the slash and the rear surface after the slash. For example, flat/flat curve is for the wafer with two flat surfaces. About 60% of the short wavelengths are reflected and, therefore, 40% of them absorbed.

At the middle of the relevant solar spectrum, reflectance drops to about 30%, absorbing the other 70%. Toward the long wavelengths, reflectance increases back to about 50%, with the rest split between absorption and transmittance.

Changes to the top surface affect the entire spectrum. Introducing regular pyramids on the top surface moves us to the textured/flat curve with much better performance than the flat/flat case, except at the longest wavelengths.

Changes to the rear surface affect only the long wavelengths that can reach it. Introducing regular 4.2 μm tall pyramids to the rear surface moves us to the textured/textured curve, which brings down reflectance above 1 μm.

On the other hand, the introduction of antireflective nitride film on the top surface dramatically reduces reflectance in the middle of the spectrum without much effect toward the ends of the solar spectrum.

Having antireflective nitride film on the rear surface reflects more infrared light toward the top than the rear surface covered by aluminum. This is important for the

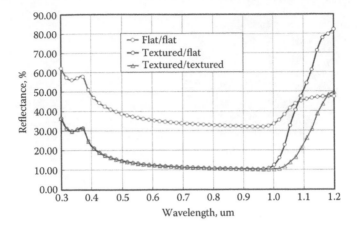

FIGURE 7.2 Calculated solar cell reflectance as a function of light wavelength. The flat/flat curve is for the silicon wafer with flat top and rear surfaces. The textured/flat curve is for the silicon wafer with textured top and flat rear surfaces. The textured/textured curve is for the silicon wafer with textured surfaces on both sides. All textured surfaces consist of regular pyramids that are 6 mm wide and 4.2 mm tall.

solar cells with point rear contacts, because the rear contacts are aluminum and the rest of the rear surface is covered with nitride. Therefore, part of the light reflects off the nitride-covered rear surface, whereas the other part reflects off the aluminum-covered rear surface. The ratio of aluminum contact area to the area of nitride-passivated rear surface determines the overall optical behavior of such solar cells.

So far, we have been analyzing regular texture with the pyramids that are facing up, which is difficult to obtain practically, but is easy to model. It is possible to manufacture regular texture with the pyramids facing down by using photolithography and wet etching. However, the need for the photolithography step makes the process too expensive for competitive manufacturing. Therefore, the industry is using wet etching without any masks, which gives random texture with the pyramids that are facing up.

7.2.3 Regular Versus Random Texture

It is more difficult to model random texture, as it involves a larger simulation domain with multiple overlapping pyramids and requires robust 3D geometry and mesh algorithms to handle such complex geometries. Due to the recent advances in mature simulation tools, such analysis is possible and its results are reported next.

When comparing the reflectance of the structure with regular pyramids to the ones with random texture, one remarkable observation is that the optical performance of random texture is noticeably better than performance of the regular texture, especially in the ultraviolet part of the solar spectrum. Let us find out why.

One hypothesis is that the random texture performs better due to the random lateral locations of the pyramids. Figure 7.3 shows a side view of the regular and random textures that exhibit very different skylines. The regular texture has rows of

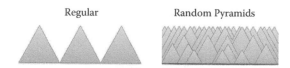

FIGURE 7.3 Sketch of the side view for regular texture on the left and random texture on the right. The two skylines are very different.

pyramids that cover exactly half of the area in the range from the foot of the pyramid to the top of the pyramid. The other half of that area belongs to the air between the pyramids.

Now, let us look at the rays that bounce out of the top surface. The rays that bounce at large angles that are close to vertical will definitely escape from the solar cell and will contribute to the wasteful reflectance. The rays with low bouncing angles that are close to the surface have a chance of being recaptured by the other pyramids. Figure 7.3 shows that about half of such rays will be captured by the regular texture, with the other half escaping.

In contrast, shifted pyramids of the random texture cover the entire skyline and can capture all of the low angle rays. This should explain the better optical performance of the random texture.

Let us perform analysis of another structure that can help to confirm this hypothesis. Specifically, let us model reflectance of the surface that has pyramids of the same height but placed at random lateral locations. Moreover, because we can control where those "random" locations are, we can keep the pyramids at the same lateral locations as we had for the true random pyramids. This ensures that there is only one variable changing at a time and the results are cleaner and easier to interpret. A structure like this would be nearly impossible to make experimentally, but is easy to model.

Figure 7.4 proves this hypothesis by confirming low reflectance of the texture with randomly placed pyramids of the same height.

Actually, performance of the artificial texture with randomly placed pyramids of the same height is about 20% higher than the performance of regular pyramids with the same height, which is quite substantial. And it is slightly higher than performance of the true random texture.

Figure 7.5 illustrates why it happens. Due to the limited size of the simulation domain with 20 μm by 20 μm surface and about 100 pyramids, the skyline of the truly random texture has some holes and, therefore, misses some of the low angle rays. In contrast, the artificial texture with randomly placed pyramids of the same height has much better skyline coverage, which explains its superior optical performance.

If we look at the structure with regular pyramids that are facing down, the skyline there is flat as in a random texture with pyramids facing up. However, the random pyramids have an advantage that most of the light scattering events happen at the bottom of the pyramids, because pyramid tips make up only a small portion of the overall surface area. The rays that are coming from the pyramid feet have a good chance of bumping into neighbor pyramids even at slightly upward ray angles.

On the other hand, in regular pyramids that are facing down, most of the surface area and, therefore, most of the scattering events happen close to the top of the

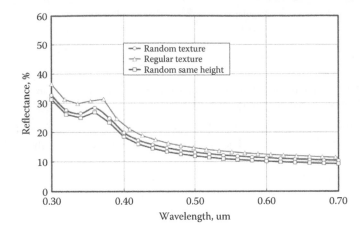

FIGURE 7.4 Reflectance of different textures as a function of light wavelength. The regular texture has 4.2 mm tall pyramids, the random texture has pyramids of random height and random lateral placement, and the "random same height" texture has pyramids that are 4.2 mm tall, but laterally placed into the same locations as the "random texture."

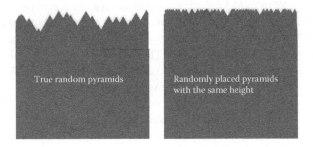

FIGURE 7.5 Side view of the simulated structures with randomly placed pyramids of random height on the left and randomly placed pyramids of the same height on the right. In both cases, simulation domain size is 20 mm × 20 mm, so we see a 20 mm wide (i.e., lateral) and 20 mm deep (in the direction perpendicular to the page) structure.

skyline. Therefore, any rays that bounce with an upward angle escape into the air, reducing the optical solar cell performance.

To summarize this section, we discussed behavior of different light wavelengths with several types of silicon surface roughness and found that the best optical performance is achieved by random texture. If the randomly placed pyramids have the same or at least similar size, it would further improve the performance, but might be difficult to manufacture.

The next section takes optically generated carriers from this section and discusses how they travel through the cell to its contacts to generate solar power.

7.3 MODELING ELECTRONIC EFFECTS

For the monocrystalline silicon cells, one appealing strategy to increase the efficiency is to introduce point contacts on the rear surface instead of the conventional structure with a contact that covers the entire rear surface.

With point contacts on the rear surface of a solar cell, several competing physical mechanisms determine its performance. On the one hand, different optical reflectivity and surface recombination rates for the silicon–aluminum interface and passivated silicon–nitride interface suggest that reducing rear contact area would boost cell efficiency. On the other hand, current crowding, contact resistance, and bulk recombination will contribute to cell performance degradation with shrinking rear contact area. Furthermore, the trade-off between these factors will be affected by any change in doping concentration, silicon quality, and cell size. It has been reported [1,4] that rear point contact can increase the open circuit voltage (Voc) and short circuit current (Jsc), at the cost of reducing cell fill factor. Therefore, there is a large optimization space to find the best solar cell design.

Optimization of the placement, including size and location, of rear point contacts is performed using 3D simulation with Sentaurus TCAD tools [3]. The simulation work flow starts from processing a precalculated optical generation profile, according to different optical reflectivity at rear surfaces with or without the contact. Sentaurus Device Editor creates a 3D structure with the processed optical profiles. Sentaurus Device performs electrical analysis of the structure to calculate illuminated currents. The results are processed to extract photovoltaic parameters like Jsc, Voc, fill factor, and power conversion efficiency.

7.3.1 DEFINITION OF SIMULATION CELL STRUCTURE

7.3.1.1 Structure Definition

The monocrystalline silicon solar cell consists of a p-type silicon substrate, front and rear surfaces covered with a passivated nitride layer, a silver front contact stripe, and rectangular aluminum rear contacts with a heavily doped region underneath the contact metal.

The solar cell structure and simulation domain boundaries are defined based on three criteria, which are illustrated in Figure 7.6 together with various geometry parameters. First, the length (along the X-axis) of the simulated element is half of the front contact pitch, with the assumption that front contact pitch is larger than rear contact pitch. Second, the width (along the Z-axis) of the simulated element is half of the rear contact pitch; therefore, only half of the rear contact will be placed along this direction. Third, the placement of the first rear contact along the X-direction is controlled by an offset parameter, and the placement of other rear contacts, if any, is decided by the rear contact size and pitch. The definitions of p–n junctions are also shown in Figure 7.6.

The number and location of rear contacts can be controlled through adjustment of the geometry parameters. The rear contact area coverage in percentage is then calculated as a measure that describes the rear contacts.

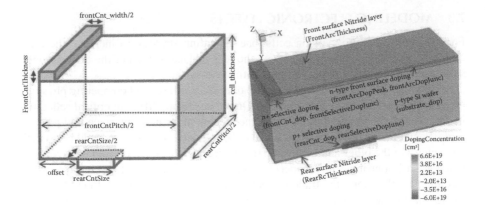

FIGURE 7.6 Definition of solar cell structure and simulation domain (left); doping profiles with p-type Si wafer, blanket n-type top surface doping, and local n+ and p+ junctions around the contacts.

7.3.1.2 Meshing Strategy

Mesh is one of the most important aspects in determining simulation efficiency and accuracy. The general practice is to apply coarse mesh to the whole region first and then to zoom into areas that require high resolution and refine the mesh in those regions. Fine mesh is necessary whenever there are material interfaces, p–n junctions, and contacts. Also, specifically for solar cells, it is expected that there be a sharp gradient of light-generated carriers within the first 30 to 40 µm from the top surface. Therefore, it is recommended to refine vertical mesh spacing toward the top surface to resolve the optically generated carrier profile.

7.3.2 Modeling Methodology

To develop an accurate simulation setup for solar cell optimization, all major physical mechanisms need to be modeled properly.

7.3.2.1 Impact of Optical Reflectivity on Optically Generated Carrier Profile

The different optical reflectivity at the rear surface for silicon–aluminum interface (i.e., with rear contact) and with passivated silicon–nitride interface (i.e., without rear contact) determines the amount of light retention within silicon and thus the number of optically generated carriers that can be trapped and absorbed in the solar cell. Because rear contact regions trap less light through reflection, it is desirable to have a smaller rear contact area from this point of view.

To simulate this mechanism, two different optical generation profiles must be placed in regions with or without the rear contact. No optically generated carriers are placed in the region underneath the front contact stripe due to the shading effect. Within each region, it is assumed that the optical profile is distributed uniformly along the horizontal plain (X–Z plain in the simulation).

7.3.2.2 Surface Recombination Rate

In addition to optical reflectivity, different interfaces also demonstrate different surface recombination rates. Because the carrier recombination velocity at the silicon–aluminum interface is much higher than that of the silicon–nitride interface, the decrease of rear contact area coverage reduces the chance of carrier recombination and improves the solar cell performance. This factor is modeled by defining different prefactors of the surface recombination velocity associated with different interfaces.

7.3.2.3 Contact Resistance

Contact resistance is affected by factors such as material properties and the dimension of the contact. For a given set of material properties, the decrease in rear contact area will increase contact resistance and degrade the solar cell performance.

7.3.2.4 Bulk Recombination

In the lightly doped silicon substrate, recombination is dominated by the defect-induced Shockley–Read–Hall recombination. In regions with high doping concentrations, such as the p–n junction at the front surface and the selective doping areas underneath contact metal, Auger recombination becomes significant. Therefore, both bulk recombination mechanisms are modeled as a function of doping concentration.

With the physical models mentioned previously and the capability to control the cell structure, we can vary the sizes of different parts of the solar cell, or depth and doping level of the junctions, or physical properties of silicon and its interfaces, and then investigate their impact on solar cell performance.

7.3.3 Current Crowding

Consider a solar cell that has 1 mm distance between top contact finger lines and 2.8% rear contact area coverage. The simulated cell power conversion efficiency is 20.93% (Jsc = 26.93 mA/cm^2 and Voc = 687 mV). Figure 7.7 demonstrates the cross-sectional view of hole current distribution at different locations.

7.3.4 Optimizing Efficiency of Solar Cells

With the other properties fixed, we vary the rear contact size to change its area coverage. Figure 7.8 illustrates relevant competing physical mechanisms that determine design trade-offs.

Figure 7.9 shows calculated cell power conversion efficiency as a function of the rear contact area coverage. The nonmonotonic trend is a result of multiple competing physical mechanisms and points toward an optimal design around 5% rear contact area coverage. Compared with the design, which has the full backside covered by rear contact, the best design offers more than 1% gain in cell efficiency.

For a different set of p–n junctions, or silicon properties, or contact pitches, the optimum design will be somewhat shifted, but can be found using the same modeling methodology. For instance, when the substrate doping level is low, the high bulk resistance causes a bigger problem in solar cells with smaller rear contact, because

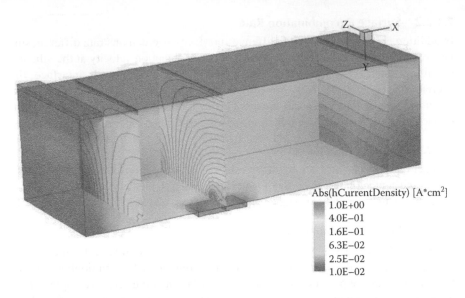

FIGURE 7.7   Slices along X-direction show hole current distribution at different cross sections.

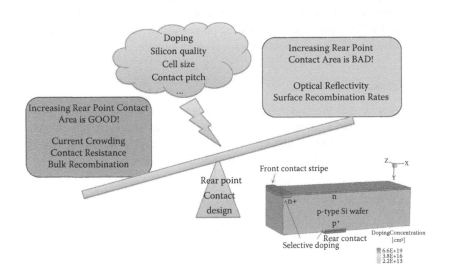

FIGURE 7.8   Design trade-offs for optimizing the size of rear point contacts.

current crowding becomes the dominating constraint of the cell efficiency. On the other hand, when substrate doping is high, the low bulk resistance is unlikely to play an important role in determining the cell performance. Therefore, small rear point contacts are desirable with high substrate doping, while the full surface rear contact is preferable when the substrate doping is low.

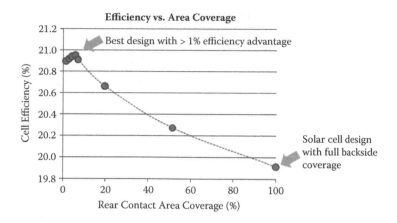

FIGURE 7.9 Efficiency of a solar cell as a function of rear contact area coverage.

7.3.5 COMPARING 3D WITH 2D AND 1D

Next we compare the three-dimensional simulation results with the simplified two-dimensional and one-dimensional counterparts. To minimize the impact of mesh-related numerical noise, the same meshing strategy is adopted in all simulations.

However, due to the nature of 1D simulation, which only allows variations along one direction (Y-axis), the structure used in 1D simulation is slightly modified by extending the front contact to cover the whole top surface and removing the heavily doped region under the front contact. The whole rear surface is also fully covered by rear contact, making it impossible to change the rear contact area coverage. Therefore, a valid comparison among 1D, 2D, and 3D simulations can only be drawn from structures with 100% rear area coverage. Two groups of comparison are performed with different parameter settings. Both results show that 1D simulation calculates cell efficiency that is about 2% lower than the 2D or 3D results, which demonstrates that 1D simulation is definitely insufficient to solve problems like this.

A comparison between 3D and 2D simulation results is shown in Figure 7.10. Noticeable discrepancies of up to 0.5% in simulated cell efficiency are observed when rear contact area coverage is small in Figure 7.10. Such discrepancies can be explained by the current crowding effect, which is captured more accurately in 3D simulations. The main geometrical difference between 3D and 2D simulation is that the rear contact has a rectangular shape in 3D, but it is a stripe of infinite length in 2D simulations. Therefore, the current crowding effect is almost doubled in 3D simulation because each contact has four corners compared to two corners in the 2D version, where current crowding takes place. According to Figure 7.10, the low substrate doping makes current crowding a dominating constraint, resulting in a bigger difference between 3D and 2D simulations.

Based on these observations, we conclude that 2D simulations are mostly good enough to produce similar results to 3D simulations, except for certain cases where we see up to 0.5% discrepancy in efficiency. However, it is recommended to use 3D simulations when current crowding effects cannot be neglected.

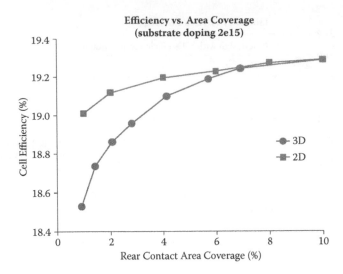

FIGURE 7.10 Three- and two-dimensional simulation comparison showing cell efficiency as a function of rear contact area coverage, with a substrate doping concentration of 2.0×10^{15} cm^{-3}.

7.3.6 JUNCTION OPTIMIZATION

Efficiency of the solar cell is almost insensitive to particular properties of the heavily doped n+ selective emitters, as long as they provide good enough conduction and low enough contact resistance. There are no optically generated carriers there because the selective emitter is in a shadow of the optically opaque silver contact on the top surface. Therefore, carrier recombination is not an issue in the emitters, so they are usually doped as heavily as possible to provide good contact resistance and good conduction.

Conversely, solar cell performance is very sensitive to the properties of blanket n+ junctions. On the one hand, higher doping in the front n-type layer can reduce resistance and, therefore, boost the cell efficiency. On the other hand, higher doping leads to higher recombination rates of minority carriers, which suppresses the cell performance.

The phosphorous oxychloride (POCL) diffusion that is used in the industry to make the n+ junctions creates surface doping of about $2 \cdot 10^{20}$ cm^{-3} ± 25%. The 50% doping range happens due to the process variations. For a typical junction depth of 0.4 µm, the efficiency is 18.8% ± 0.25%, as can be seen on Figure 7.11.

Figure 7.11 also shows that a better junction with peak doping of 10^{19} cm^{-3} and junction depth of 0.6 µm can simultaneously boost the efficiency by 2.25% and significantly reduce the efficiency variability, from 0.5% down to 0.05%. Reduction of variability tightens the efficiency spread and, therefore, improves parametric yield. In terms of process flow, such junctions can be obtained by adding anneal with large thermal budget to the standard POCL process to reduce the surface doping, or to etch

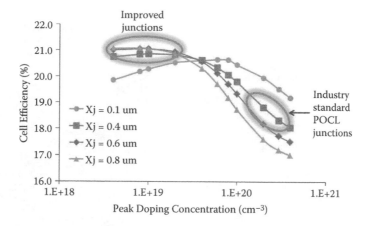

FIGURE 7.11 Comparative performance and variability of different n+ blanket junctions.

away heavily doped surface layer, or to use alternative doping techniques such as ion implantation or plasma doping.

Another option is to keep the conventional POCL doping process, but make shallower junctions of about 0.1 μm deep. This will increase efficiency by about 1%, but will not significantly reduce the variability.

To summarize this section, we discussed several design optimization criteria and found optimization space of 4% in terms of efficiency for the junction design, 1% for the rear contact size, and 4% for the substrate doping. Additionally, we found that 1D modeling is not accurate enough, but 2D modeling gives reasonable results most of the time, with the maximum observed discrepancy with 3D modeling of 0.5% in terms of cell efficiency.

7.3.7 An Alternative Solar Cell Design

Among efforts to boost the performance of monocrystal silicon solar cells, the all-back-contact cell design is regarded as an appealing candidate. Moving all contacts to the back surface and leaving no metallization pattern on the front surface eliminates the optical shading losses and allows the front surface to be optimized without the trade-off between grid shading and series resistance. It is projected that the worldwide fraction of all-back-contact solar cells in production will increase to 35% by 2020 from its current share of 5% [5].

For an all-back-contact solar cell, many options remain to be explored in order to harvest the optimal cell efficiency. Take the front surface doping as an example for which there are at least two approaches. Assuming the same lightly doped n-type substrate, one way is to have a heavily doped n+ layer at the front surface, which reduces the series resistance and provides a passivating drift field. The alternative is a heavily doped p+ layer, which serves as a floating emitter and provides an even larger drift field, but it needs careful design to minimize carrier recombination in the heavily doped surface layer. It is therefore useful to explore the possibilities in the all-back-contact cell design through TCAD simulations.

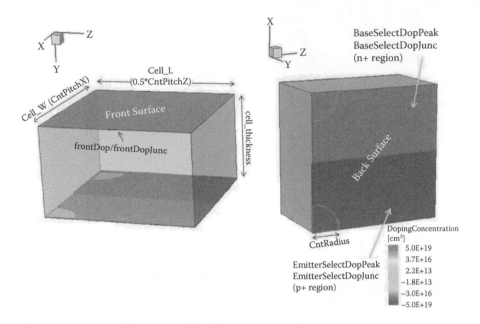

FIGURE 7.12   Illustrations of various input parameters defining the geometry and doping in the simulation structure. The oxide layer covering the back surface is not shown.

The comparison among various design options is performed using 3D simulation with Sentaurus TCAD tools, with a simulation work flow and modeling methodologies similar to the point rear contact example described previously.

7.3.7.1   Structure Definition

The nominal structure is illustrated in Figure 7.12. The silicon substrate is lightly doped with phosphorus to a concentration of 10^{15} cm^{-3}, and the front surface bears a heavily doped n+ layer with the peak doping of 6×10^{19} cm^{-3}. Alternate n+ and p+ stripes are placed on the back surface, with a round contact in each stripe. A nitride antireflecting layer covers the front surface, and the back surface outside the contacts is covered by oxide.

Four types of variations in the cell design are considered in the simulation. These include the front surface doping type, the contact pitch, the substrate doping level, and changes to localized back surface doping.

7.3.7.2   Results and Discussion

Table 7.1 summarizes the impacts on cell performance due to each variable considered. It can be concluded that, for this particular setup, having a p+ junction at the front surface can help increase cell performance when compared with the other two alternative front surface treatments, which are n+ doped and no heavily doped surface.

The cell efficiency of structures is directly related to the area of the heavily doped regions on the back surface; therefore, structures with localized back surface doping are not as good as the ones with n+ and p+ doping stripes.

TABLE 7.1

Summary of Main Simulation Results

Variable	Substrate Doping (cm⁻³)	Contact Style	Front Doping (cm⁻³)	Jsc (mA/cm²)	Voc (mV)	FF (%)	Eff. (%)
Nominal	$1E + 15$	Stripe	$6E + 19$	34.12	642.9	80.71	20.78
Floating front junction	$1E + 15$	Stripe	$-6E + 19$	38.47	653.2	80.57	23.77
Nonheavily doped front surface	$1E + 15$	Stripe	$1E + 16$	37.09	621.9	79.88	21.63
Localized back surface doping	$1E + 15$	Circular (R = 80 μm)	$6E + 19$	29.71	638.1	78.67	17.51
	$1E + 15$	Circular (R = 170 μm)	$6E + 19$	34.13	641.0	79.91	20.52
Substrate doping	$1E + 14$	Stripe	$6E + 19$	35.03	643.9	79.34	21.01
	$1E + 16$	Stripe	$6E + 19$	28.91	641.1	82.94	18.04
Contact pitch	$1E + 15$	Stripe	$6E + 19$	34.16	643.4	80.00	20.64
	$1E + 15$	Stripe	$6E + 19$	33.43	643.4	80.14	20.24

There are competing factors associated with changes in substrate doping. When the doping level in silicon substrate increases, the carrier lifetime decreases, which leads to lower conducting current. At the same time, the resistance is reduced, which is beneficial to the cell efficiency.

With the bulk carrier lifetime of 500 μs assumed in this simulation, an increase in the pitch between n-contact and p-contact from 250 to 500 μm leads to a 0.5% drop in cell efficiency.

Due to the complex mechanisms involved in the all-back-contact solar cell design, if any of the conditions is modified, the design trade-offs change, requiring reoptimization.

7.4 CONCLUSIONS

We discussed design of optical and electrical aspects of silicon solar cells using 3D simulation with comprehensive physical models. Simulation results reveal significant impact of surface texture and antireflective layers on sunlight capture. Robust mesh and geometry building tools enable analysis of large simulation domains with random texture. Simulations with about 100 random pyramids were large enough to characterize random texture reproducibly. Detailed comparison of regular and random textures reveals the exact reasons behind better optical performance of the cell with random texture.

Electrical analysis of the solar cell with rear contact covering anywhere from 100% down to 1% area reveals significant optimization space of over 4% in terms of efficiency. The large number of competing physical mechanisms leads to complex cell behavior with the trade-off points determined by a combination of several design parameters.

The performed optical and electrical analyses suggest several possible ways to improve cell performance and tighten its variability.

REFERENCES

1. D. Kray, N. Bay, G. Cimiotti, et al., 2010. Industrial LCP selective emitter solar cells with plated contacts. Photovoltaic Specialists Conference.
2. S. W. Glunz, J. Knobloch, C. Hebling, and W. Wettling. 1997. The range of high-efficiency silicon solar cells fabricated at Fraunhofer Ise. Photovoltaic Specialists Conference.
3. Sentaurus TCAD tools, v. 2013.03, Synopsys, 2013.
4. M. Green, and A. Blakers 1990. Characterization of 23% efficient silicon solar cells. *IEEE Transactions on Electron Devices* 37 (2): 331–336.
5. International Technology Roadmap for Photovoltaics (ITRPV) Results 2011. Available at: <http://www.itrpv.net/doc/roadmap_itrpv_2012_full_web.pdf>

8 Piezoelectric Thin Films and Their Application to Vibration Energy Harvesters

Isaku Kanno

CONTENTS

8.1 INTRODUCTION

8.1.1 PIEZOELECTRIC MEMS

Microelectromechanical systems (MEMS) technologies have attracted considerable attention for the development of next-generation functional microdevices. In general, MEMS devices' functionality is originated from three-dimensional microstructures that can be fabricated by Si microfabrication technologies; therefore, Si-based materials are mainly used. If we think about microactuators, driving force is generally categorized into four systems: electrostatic, electromagnetic, thermal, and piezoelectric.

Electrostatic	Magnetic	Thermal	Piezoelectric
$F(x) = \dfrac{1}{2}\dfrac{\varepsilon_0 S}{(d-x)^2}V^2$	$F = Bil$	$F/S = \alpha \cdot \Delta T \cdot E$	$F_1/S = E \cdot d_{31} \cdot \dfrac{V}{t}\ (i=1,3)$
Easy microfabrication Low power consumption Fast response	Conventional and traditional actuators Remote operation	Large force Simple structure	Fast response Low voltage Large force
Small attractive force Limited stroke High voltage	Microfabrication Generation of heat	Cross talk Slow response	Thin film growth Microfabrication

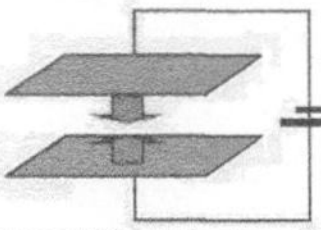

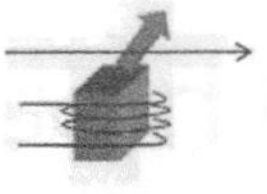

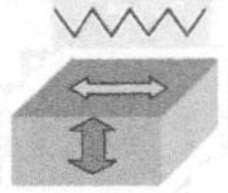

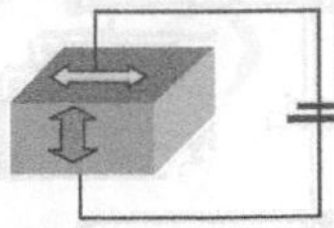

FIGURE 8.1 Driving force of MEMS actuators.

Each system has advantages and disadvantages (Figure 8.1). The most popular MEMS actuators are electrostatic systems, which are composed of simple materials such as Si microstructure and metal electrodes. Furthermore, design and microfabrication of Si-based MEMS have already been established in semiconductor microfabrication technologies. However, generated force of electrostatic actuators is only attraction, and the stroke is limited in gap distance. On the other hand, functional materials such as piezoelectric materials, which have gradually been integrated into MEMS, can give new functionality on simple microstructures.

Piezoelectricity originated from the charge asymmetry in dielectric crystals. Deformation of a piezoelectric material generates an electric dipole in a unit cell of the crystal, and summation of individual dipoles becomes polarization, which is converted to the electric output (Figure 8.2). Piezoelectricity has two characteristics: One is the piezoelectric effect, which means the charge generation by external stress or strain, and the other is the inverse piezoelectric effect, which is force generation by an external electric field. These characteristics imply that the piezoelectric materials are inherently sensors and actuators. Therefore, if we integrate piezoelectric

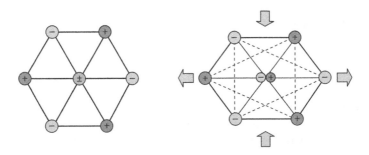

FIGURE 8.2 Deformation of a piezoelectric crystal. Asymmetry of electric charge causes polarization and electric output.

materials into MEMS, microsensors or actuators are easily developed, especially in simple microstructures.

8.1.2 PIEZOELECTRIC THIN FILMS

For integration of piezoelectric materials into MEMS, the piezoelectric materials should be prepared in thin-film form. The most popular piezoelectric material is $Pb(Zr,Ti)O_3$ (PZT) because of its excellent energy conversion efficiency between mechanical and electrical domains. The deposition and characterization of PZT thin films have been intensively studied; recently, some piezoelectric MEMS composed of PZT films have been developed as commercial products.

In the case of energy harvesters, the piezoelectric property is one of the most important factors to determine power generation efficiency from environmental vibration, and PZT thin films are commonly applied to small-scale vibration generators. However, issues still remain: deposition of the well-crystallized piezoelectric films with complex composition and precise measurement of piezoelectric properties of thin films. In addition, PZT contains toxic lead, which would be an issue for the environment and human health.

In this chapter, the deposition process of PZT thin films is described from the viewpoint of MEMS applications. In addition, the recent progress of the lead-free piezoelectric thin films is introduced. Among a variety of piezoelectric MEMS, we pick up the piezoelectric MEMS energy harvesters, and the basic characteristics and applications of piezoelectric MEMS energy harvesters are presented.

8.2 PREPARATION OF PIEZOELECTRIC PZT THIN FILMS

8.2.1 DEPOSITION OF PZT THIN FILMS FOR MEMS

For the deposition of the piezoelectric materials, especially PZT-based perovskite thin films, several deposition techniques have been studied: chemical solution deposition (CSD; sol-gel deposition) [1], chemical vapor deposition (CVD) [2], pulsed laser deposition (PLD) [3], and sputtering [4]. These four methods can produce well-crystallized PZT thin films; however, each of them has advantages and disadvantages from the viewpoint of piezoelectric MEMS, as shown in Table 8.1. Among these methods, CSD is one of the popular techniques to prepare piezoelectric PZT thin films because it does not need expensive deposition apparatus. However, multiple

TABLE 8.1

PZT Deposition Process in MEMS

Process	Advantage	Disadvantage
CSD	Low cost	Multiple coating
CVD	Uniform coverage	Cost of source material
PLD	Reproducibility of target composition	Small area
Sputter	Fast deposition rate	Cost of apparatus

spin-coating and baking processes are necessary to obtain a thickness of more than 1 μm, which often causes composition variation along the thickness and reduction of piezoelectric properties [5].

On the other hand, CVD is another candidate for high-quality PZT deposition. The advantage of this method is excellent step coverage, which is useful for semiconductor memories (FeRAM). However, because the source materials of PZT are expensive compared to the other methods, it is not suitable for the deposition of micron-thick piezoelectric PZT films in piezoelectric MEMS. PLD is also a popular method for PZT deposition; however, the deposition area with uniform thickness in a Si wafer is usually small and it is difficult to fabricate piezoelectric MEMS devices as commercial products.

For the PZT deposition in piezoelectric MEMS, radio frequency (RF)-magnetron sputtering is a practical method to produce micron-thick PZT films in a single deposition at lower cost. Sputtering deposition enables PZT films to be formed on a variety of substrates with high deposition rates of more than 1 μm/h. On the other hand, sputtering deposition synthesizes the films in a nonthermal equilibrium condition; the deposited PZT thin films show unique characteristics that the PZT thin films of CSD or CVD do not have. For example, sputtered PZT thin films usually contain excess Pb, which is located not in A site but in B site [6,7]. Furthermore, the sputtered PZT thin films have stable initial poling from bottom to surface, and this phenomenon is useful from the viewpoint of long-term duration of the piezoelectric MEMS sensors [4,8]. Sputtering deposition of the piezoelectric PZT thin films has already been used in mass production, such as ink-jet printer heads and gyro sensors [9].

8.2.2 SPUTTERING DEPOSITION

The RF-magnetron sputtering is a useful deposition technique, especially for the thin films composed of complex chemical composition. Typical sputtering conditions of the PZT thin films are listed in Table 8.2. The sputtering is performed under an Ar/O_2 mixed gas atmosphere of 0.3 ~ 0.5 Pa, where the substrates were heated up to around 600°C to grow the PZT films with perovskite structure. Postannealing with the temperature around 600°C is also effective to crystallize PZT films deposited at low substrate temperature. However, postannealing often involves fatal damage,

---

TABLE 8.2
Sputtering Condition of PZT

Target	$[Pb(Zr_{0.53}Ti_{0.47})O_3]_{0.8} + [PbO]_{0.2}$
Film composition	Zr/Ti = 53/47
Substrates	(100)Pt/MgO, (111)Pt/Ti/Si
Deposition temperature	~600°C
Sputtering gas	$Ar/O_2 = 9/1 ~ 19/1$
Pressure	0.3 ~ 0.5 Pa
Film thickness	2 ~ 3 μm

such as cracking or peeling of the films, especially for the thick PZT films due to large thermal stress from the substrate.

Sintered ceramics are usually used as target materials to stabilize film composition. It is well known that the morphotropic phase boundary (MPB) exists at a Zr:Ti ratio of 52:48, where dielectric and piezoelectric properties exhibit maxima. Therefore, the PZT ceramic target with MPB composition is usually used for the piezoelectric PZT films. Furthermore, since Pb and PbO are easily re-evaporated from the deposited films, excess PbO was often added in the target to maintain stoichiometory of the PZT films. Stabilization of Pb is an important factor to obtain the PZT thin films with perovskite structure, and depletion of Pb usually causes a pyrochlore phase in the deposited films. On the other hand, excess Pb is thought to be settled in B site of the perovskite, and it does not cause the serious degradation of the electric properties of the sputtered PZT thin films [6,7].

8.2.3   CRYSTAL STRUCTURE OF PZT THIN FILMS

Piezoelectricity is originated from asymmetry of the crystal structure, which strongly influences piezoelectric properties. Usually, PZT films are grown on Pt-coated Si substrates with a Ti adhesive layer at the interface of Si. The x-ray diffraction (XRD) patterns of the PZT films deposited on Pt/Ti/SiO$_2$/Si substrates are shown in Figure 8.3. The clear diffraction peaks from perovskite PZT is observed without the other phases such as pyrochlore. On the other hand, the preferential orientation of the PZT can be controlled by optimization of the sputtering conditions. Especially, a seed layer such as (Pb, La)TiO$_3$ (PLT) or SrRuO$_3$ (SRO) is effective in ensuing perovskite growth and orientation control in PZT sputtering [10–12].

The preferential orientation of the piezoelectric thin films is one of the determining factors of the piezoelectric properties for piezoelectric thin films. Although PZT thin films on Si substrates usually have polycrystalline structure, epitaxial substrates such as MgO and SrTiO$_3$ enable epitaxial growth of PZT films by RF-sputtering, PLD, or CVD. Epitaxial substrates can control orientation and domain structure of the PZT films so that the electric properties including piezoelectricity can be

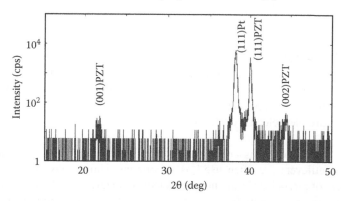

FIGURE 8.3   XRD patterns of PZT thin films on Pt/Ti/Si substrates by RF-magnetron sputtering.

enhanced by perfect alignment of the polar direction. Recently, Pb-based piezo-electric thin films have been grown on Si substrates by using a multilayered buffer such as $SrTiO_3/MgO/TiN$ or $SrRuO_3/SrTiO_3$. Baek et al. reported that $Pb(Mg_{1/3}Nb_{2/3})O_3$-$PbTiO_3$ (PMN-PT) thin films were grown on miscut Si substrates with multi-buffer layers by off-axis sputtering deposition, and large piezoelectric properties could be obtained [13]. The epitaxial piezoelectric thin films of PZT or PMN-PT are ideal materials for piezoelectric MEMS; however, the sputtering conditions are very sensitive to the epitaxial growth of the piezoelectric films including buffer layers, and the polycrystalline thin films with preferred c-axis orientation are usually used for the practical applications.

8.2.4 EVALUATION OF PIEZOELECTRIC PROPERTIES FOR THIN FILMS

Since the piezoelectric films are deposited on substrates, the piezoelectric effect of the thin film is strongly suppressed by the existence of the substrate, and it makes it difficult to measure the precise piezoelectric properties. Several measurement methods have been used to determine piezoelectric coefficient of the thin films [14]. One of the most popular methods is the measurement of the thickness change originated from the longitudinal piezoelectric effect using an atomic force microscope (AFM) or laser interferometer [15,16]. Although this method offers ease of measurement of the effective longitudinal piezoelectric strain, the transverse piezoelectric coefficient, which is indispensable for the device design of piezoelectric MEMS, cannot be obtained.

The most practical method to evaluate transverse piezoelectric properties of thin films is the measurements of piezoelectric cantilevers. The piezoelectric properties of d- and e-forms are expressed as the following equations:

$$T_i = c_{ij}^E S_j - e_{ki} E_k, \qquad (8.1)$$

$$S_i = s_{ij}^E T_j - d_{ki} E_k, \qquad (8.2)$$

where
  $k = 1, 2, 3$
  $i, j = 1, 2, 3, 4, 5, 6$
  T_i, S_j, and E_k are the stress, strain, and electric field components
  c_{ij}^E and s_{ij}^E are constant-electric-field elastic stiffness components and elastic compliance components
  e_{ki} and d_{ki} are piezoelectric coefficients

In order to evaluate transverse piezoelectric properties (d_{31} or e_{31}) of thin films, unimorph cantilevers are often used as specimens. Unimorphs are composed of double layers of piezoelectric and nonpiezoelectric materials, and most piezoelectric MEMS sensors or actuators have these structure. Output of a unimorph is determined by transverse piezoelectric properties; for example, tip displacement of the unimorph actuator is calculated by the following equation:

$$\delta = -\frac{3s_{11,p}^E s_{11,s} h_s \left(h_s + h_p\right) V L^2 d_{31}}{K},$$

$$K = 4s_{11,p}^E s_{11,s} h_s \left(h_p\right)^3 + 4s_{11,p}^E s_{11,s} \left(h_s\right)^3 h_p + \left(s_{11,p}^E\right)^2 \left(h_s\right)^4 \tag{8.3}$$

$$+\left(s_{11,s}^E\right)^2 \left(h_p\right)^4 + 6s_{11,p}^E s_{11,s} \left(h_s\right)^2 \left(h_p\right)^2.$$

Here, h, s_{11}, L, V, and δ are the thickness, the elastic compliance, the length of the can-tilever, the applied voltage, and the tip displacement, respectively [17]. The subscripts s and p denote the substrate and the PZT thin film, respectively. If the dimension and mechanical properties of the unimorphs are known, the transverse piezoelectric coefficient d_{31} is obtained from Equation (8.3). Especially in the case in which the thickness of a substrate is much larger than that of a piezoelectric thin film, trans-verse piezoelectric coefficient d_{31} can be approximated as

$$d_{31} \cong -\frac{h_s^2}{3L^2} \frac{s_{11,p}^E}{s_{11,s}} \frac{\delta}{V}. \tag{8.4}$$

The mechanical properties of the thin films usually depend on the deposition process and the substrate. Therefore, it is inevitable to measure the elastic proper-ties of the PZT thin films precisely to obtain transverse piezoelectric coefficient d_{31}, although they are difficult to evaluate.

On the other hand, in the case of a piezoelectric thin film on the substrate, the effects of plane and shear strains can be neglected. If an external electric field is applied to a unimorph actuator, two-dimensional in-plain strain is generated and the transverse piezoelectric coefficient e_{31} is written as

$$e_{31} = \frac{c_{13}^E S_3 - T_1}{E_3} = \frac{c_{13}^E S_3}{E_3} + \frac{d_{31}}{s_{11}^E + s_{12}^E}. \tag{8.5}$$

The effective transverse piezoelectric coefficient $e_{31,f}$ is defined as $e_{31,f} = d_{31}/(s_{11}^E + s_{12}^E)$ [18], and then Equation (8.5) is expressed as

$$e_{31} = e_{31,f} + \frac{c_{13}^E}{c_{33}^E} e_{33}. \tag{8.6}$$

The effective transverse piezoelectric coefficient was defined to eliminate the ambiguous mechanical properties of the thin films, and the following coefficient can be determined from the displacement of a unimorph cantilever:

$$e_{31}^* = \frac{d_{31}}{s_{11,p}^E} \cong -\frac{E_s h_s^2}{3VL^2} \delta. \tag{8.7}$$

$$e_{31,f} = \frac{d_{31}}{s_{11,p}^E + s_{12,p}^E} \cong -\frac{h_s^2}{3\left(s_{11,s} + s_{12,s}\right)L^2V}\delta = -\frac{E_s h_s^2}{3\left(1 - v_s\right)VL^2}\delta. \qquad (8.8)$$

Here, E_s, v_s are Young's modulus and the Poisson ratio of substrate [19,20]. The effective transverse piezoelectric coefficient $e_{31,f}$ or e_{31}^* can be easily obtained from the tip displacement of a unimorph cantilever and it is useful to evaluate the transverse piezoelectric property as a kind of figure of merit.

8.3　LEAD-FREE PIEZOELECTRIC THIN FILMS

From the viewpoint of environmental impact of piezoelectric materials and devices, lead-free piezoelectric materials have been strongly demanded. Recently, a variety of lead-free piezoelectric materials has been studied and some of them show large piezoelectric properties comparable with PZT ceramics [21,22]. Typical lead-free piezoelectric materials are listed in Table 8.3. The studies on Ba, Bi, and alkali-based perovskite materials exhibited relatively large piezoelectric properties. Recently, it has been reported that the (K_1Na) NbO_3 (KNN)-based piezoelectric ceramics, whose crystal orientation and additives are optimized, show a large piezoelectric coefficient and high Curie temperature that are comparable with PZT [26]. This discovery boosted the research of lead-free piezoelectrics.

In order to integrate lead-free piezoelectric materials into MEMS, a variety of lead-free piezoelectric ceramics has been prepared on Si substrates in the thin-film form. The most promising lead-free piezoelectric thin films are KNN thin films. Just like PZT thin films, KNN thin films can be deposited by several deposition processes, such as RF-sputtering [27,28], CVD [29], and PLD [30,31]. In the case of KNN thin films, RF-magnetron sputtering can produce high-quality thin films and the sputtering conditions are almost the same as those of PZT. Shibata et al. reported successfully fabricating the KNN thin films by RF-magnetron sputtering and confirmed excellent piezoelectric properties compatible with PZT [32,33].

TABLE 8.3
Piezoelectric Properties of Typical Lead-Free Piezoelectric Ceramics

System	Composition	d_{33} (pC/N)	ε_r	Tc (°C)	Ref.
BT–BKT	$0.95BaTiO_3 - 0.05(Bi_{1/2}K_{1/2})TiO_3 + Mn(0.1\ we\%)$	76	779	168	21
BT–BNT	$0.06BaTiO_3 - 0.94(Bi_{1/2}Na_{1/2})TiO_3$	125	580	288	23
KNN	$(K_{0.5}Na_{0.5})NbO_3$	80	430	420	24
<100>–KNN	$(K_{0.44}Na_{0.52}Li_{0.04})(Nb_{0.86}Ta_{0.10}Sb_{0.04})O_3$	416	1570	253	25
BCT–BZT	$0.5Ba(Zr_{0.2}Ti_{0.8})TiO_3 - 0.5(Ba_{0.7}Ca_{0.3})TiO_3$	620	3060	93	26

8.3.1 Piezoelectric Properties of KNN Thin Films

The transverse piezoelectric properties were evaluated from the tip displacement of unimorph cantilevers, and the transverse piezoelectric coefficient $e_{31,f}$ is calculated by Equation (8.8). To examine the piezoelectric characteristics of the sputtering-deposited KNN thin films, $e_{31,f}$ was compared with that of PZT thin films. First, the KNN and PZT thin films were deposited on Si substrates (200 μm). The thickness of the KNN and PZT thin films was 3 and 2 μm, respectively. The film composition of both films was close to MPB composition (KNN: K/Na = 45/55, PZT: Zr/Ti = 50/50). XRD patterns of the KNN and PZT thin films are shown in Figure 8.4. The KNN film shows a preferential orientation of (001) in a perovskite unit cell. On the other hand, the PZT film shows relatively random orientation with strong diffractions of (001) and (110). The cross-sectional SEM (scanning electron microscopy) images of KNN and PZT films are shown in Figure 8.5. Both films have very dense structure without pores or cavities at the interface of the substrates, implying that the films strongly adhered to the substrates.

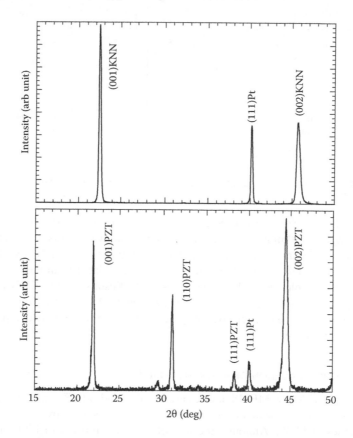

FIGURE 8.4 XRD patterns of KNN and PZT thin films deposited on Pt/Ti/Si substrates.

FIGURE 8.5   Cross-sectional SEM images of KNN and PZT thin films deposited on Pt/Ti/Si substrates.

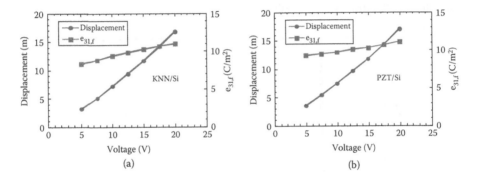

FIGURE 8.6   Displacement and calculated $e_{31,f}$ as a function of applied voltage; (a) KNN/Si and (b) PZT/Si.

The KNN and PZT films on Si substrates were cut into beam shape (width of 3.4 mm, length of 20 mm), and one end of each beam was clamped in a vise so as to create a simple unimorph cantilever (length of 16.5 mm for KNN/Si, 17.5 mm for PZT/Si). The piezoelectric coefficient $e_{31,f}$ was calculated from the tip displacement by applied voltages in the range of 5–20 V. Results for the KNN and PZT films, respectively, are $e_{31,f} = -8.4$ to -11.0 C/m^2 and -9.3 to -11.1 C/m^2, as shown in Figure 8.6. Thus, the piezoelectric properties of KNN thin film are comparable to those of PZT thin film.

8.3.2   MICROFABRICATION OF KNN THIN FILMS

At present, piezoelectric MEMS devices are usually made of PZT thin films, and the PZT thin films are fabricated into a variety of microstructures by wet or dry etching processes [4,34,35]. Although the microfabrication process of PZT-based thin films has been established, there are few reports on the microfabrication of lead-free piezoelectric thin films. In the case of KNN thin films, effective wet- and dry-etching methods have not been established so far because KNN is stable for acid or alkaline. Kurokawa et al. reported that the KNN thin film was microfabricated into a microcantilever by reactive ion etching (RIE) with Ar/C_4F_8 mixed gas and evaluated etching characteristics of KNN thin film [36]. Microfabrication of KNN thin film was conducted by magnetic neutral loop discharge (NLD) RIE, which could produce uniform and high-density plasma at low pressure and low electron temperature [15]. The etching gas flow rate of Ar/C_4F_8 was 50/2.5 standard cubic centimeters per minute (SCCM) under the pressure of 0.15 Pa; the antenna power of 800 W and bias power of 250 W were applied for 35 min. A Cr metal mask was used because conventional photoresist could not sustain high-density plasma etching. Figure 8.7 shows the KNN thin films on a Si substrate after NLD-RIE etching. KNN thin films were successfully etched into a cantilever-shaped structure. Then, cantilevers were released by Si deep-RIE (DRIE), and the SEM image of the resulting KNN microcantilevers is shown in Figure 8.8. Although the cantilevers were strongly bent by the asymmetric internal stress of KNN thin films and electrodes, they showed clear ferroelectric and piezoelectric properties.

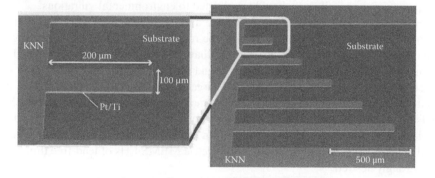

FIGURE 8.7　SEM image of microfabricated KNN thin films on Pt/Ti/Si substrate.

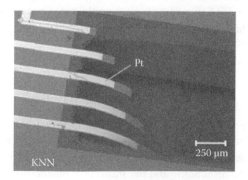

FIGURE 8.8　SEM image of KNN microcantilevers released from Si substrate by DRIE.

8.4 VIBRATION ENERGY HARVESTERS

8.4.1 PIEZOELECTRIC MEMS ENERGY HARVESTERS

Recently, vibration energy harvesting technologies have attracted considerable attention as an energy source for microdevices, especially for autonomous wireless sensor nodes [37–39]. Vibration MEMS energy harvesters (EHs) can be long-life, maintenance-free electric power sources to replace primary batteries. As vibrational energy widely exists in the environment, it is therefore useful in areas where solar power cannot be used. It is reported that the sensor nodes can be operated by small power of about tens to hundreds of microwatts [6]; however, the generated power is still insufficient and improvements in efficiency are strongly demanded.

Vibration generators are generally classified into three types: electromagnetic, electrostatic, and piezoelectric systems. Among them, piezoelectric power generators show the highest power density [37]. Furthermore, because piezoelectric energy harvesters have simple cantilever structures, integration with MEMS processing is not difficult. Piezoelectric MEMS EHs are usually unimorph cantilever structures constructed from piezoelectric PZT thin films on Si substrates. Because the vibration amplitude of the cantilever is maximized at its resonance frequency, the resonance frequency of the cantilevers should be adjusted to environmental vibrations.

However, conventional MEMS EHs have a higher resonance frequency of more than 1 kHz because of the small size of the resonators, although the frequency of environmental vibrations is generally less than 200 Hz [40]. Furthermore, PZT/Si unimorph cantilevers are brittle materials and the output power is restricted by the fracture strength of the cantilever under large vibration. Another challenge is the lead-free piezoelectric energy harvesters. Among a variety of piezoelectric MEMS EHs, PZT thin films are the most popular piezoelectric materials. On the other hand, as described in a previous section, KNN thin films have been investigated for use in lead-free piezoelectric MEMS, and piezoelectric MEMS EHs of lead-free thin films are strongly desired.

Piezoelectric MEMS EHs of piezoelectric thin films are described in this chapter. First, the power generation performance of simple lead-free KNN thin-film EHs is compared with the PZT thin-film EHs. Next, we present the metal-based piezoelectric MEMS EHs. The PZT thin films were directly deposited on the microfabricated stainless steel cantilevers by RF-magnetron sputtering and fabricated vibration EHs not only for a simple, low-cost fabrication, but also for a long lifetime of robust piezoelectric cantilevers.

8.4.2 POWER GENERATION OF PZT AND KNN THIN FILMS

Power generation performances of KNN and PZT thin films were evaluated. To simplify the characterization, simple unimorph cantilevers of KNN and PZT thin films without seismic mass were used for measurements. The samples were the same cantilevers used in Section 8.3.1, and the transverse piezoelectric coefficient $e_{31,f}$ of KNN and PZT thin films was around -10 C/m^2. The relative dielectric constants (ε_r) of the KNN and PZT films were 744 and 872, and the dielectric losses (tan δ) were

7.1% and 3.9%, respectively. The power generation efficiency of the piezoelectric thin films can be evaluated by the figure of merit (FOM) from the following equation [13]:

$$FOM = \frac{\left(e_{31,f}\right)^2}{\varepsilon_r \varepsilon_0}. \tag{8.9}$$

FOM represents the electromechanical coupling factor of the thin film materials. Because piezoelectric and dielectric properties of the KNN and PZT films are almost the same, their FOM is calculated to be around 15 GPa. This result indicates that the lead-free KNN thin films have a comparable power-generation performance to that of PZT thin films.

Output power of each unimorph cantilever was measured using the setup shown in Figure 8.9. The cantilever was mounted on a shaker and connected lead lines between the top and bottom electrodes to load resistance. Then, the output voltage was measured under the continuous vibration. The frequency response of output voltage is shown in Figure 8.10. The output voltage (for acceleration 10 m/s², load resistance $R = 1$ MΩ) showed clear peaks at frequencies of 1036 and 892 Hz and

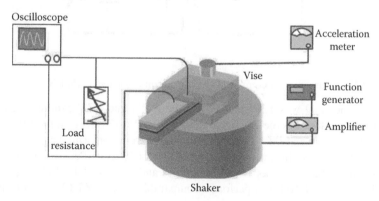

FIGURE 8.9   Measurement setup of piezoelectric energy harvesters.

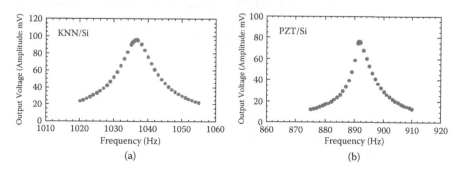

FIGURE 8.10   Frequency response of output voltage; (a) KNN/Si, and (b) PZT/Si unimorph EHs.

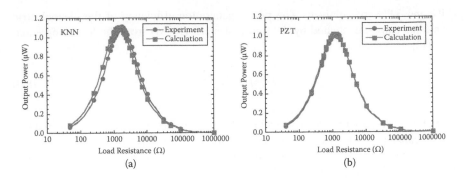

FIGURE 8.11   Output power as a function of load resistance; (a) KNN/Si, and (b) PZT/Si unimorph EHs.

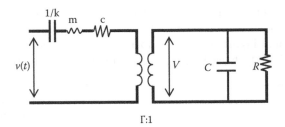

FIGURE 8.12   Illustration of equivalent electric circuit of piezoelectric EH.

the quality factor Q was 120 and 160 for the KNN and PZT cantilevers, respectively. Figure 8.11 shows the average output power $P = V^2/(2R)$ as a function of load resistance for KNN and PZT EHs. Measurements were performed at the resonance frequency of each cantilever and the acceleration of 10 m/s². Peak output power for the KNN and PZT films was 1.1 μW at 1.7 kΩ and 1.0 μW at 1.2 kΩ, respectively. Thus, the lead-free KNN film performs comparably to the PZT film with respect to power generation.

The power generation for the piezoelectric EHs can be characterized by considering the equivalent electric circuit [41]. The mechanical and electrical circuit model is illustrated in Figure 8.12. The mechanical vibration of the EHs is simplified in a lumped model, as shown in Figure 8.13, and the motion equation is

$$m\ddot{z} + c\dot{z} + kz = -m\ddot{y}, \tag{8.10}$$

where m, c, and k are mass, damping coefficient, and spring constant, and z and y are relative displacement of mass and the displacement of the base. If the conversion ratio between electric potential (V) and mechanical force (F) is Γ, Equation (8.10) can be expressed as

$$F(t) = -m\ddot{y} = -m\left(Y_0 \sin \omega t\right)'' = mY_0\omega^2 \sin \omega t = \left(\frac{k}{j\omega} + j\omega m + c\right)\dot{z} + \Gamma V. \tag{8.11}$$

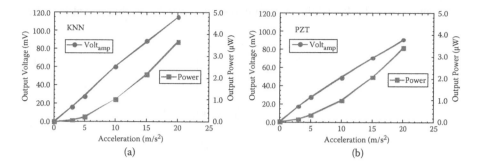

FIGURE 8.13 Output voltage and electric power of (a) KNN and (b) PZT EHs as a function of acceleration.

Then, the output voltage is expressed as the following equation:

$$V = \frac{jmY_0\omega^3 R\Gamma}{k + j\omega(c + kCR + R\Gamma^2) - \omega^2(m + cCR) - j\omega^3 mCR} \sin\omega t$$

$$= \frac{m\omega_0^2 Y_0}{\Gamma} \frac{K^2\Psi\Omega^3}{j - \{(1/Q) + (1 + K^2)\Psi\}\Omega - j(1 + (\Psi/Q))\Omega^2 + \Psi\Omega^3} \sin\omega t$$

(8.12)

$$\omega_0 = \sqrt{\frac{k}{m}} \quad \omega_e = \frac{1}{RC} \quad \Psi = \frac{\omega_0}{\omega_e} \quad \Omega = \frac{\omega}{\omega_0} \quad Q = \frac{m\omega_0}{c} \quad K^2 = \frac{\Gamma^2}{kC},$$

where m, a, ω_0, C, R, and Q are the cantilever's effective mass, acceleration, fundamental mechanical angular frequency, capacitance, load resistance, and mechanical quality factor. K represents the generalized electromechanical coupling (GEMC) factor. The averaged output power is obtained as

$$P = \frac{1}{T}\int_0^T \frac{V^2}{R} dt$$

(8.13)

$$= \frac{mY_0^2\omega^3}{2} \frac{K^2\Psi\Omega^6}{(\Omega^2 - 1)^2 + \Psi^2\Omega^2(\Omega^2 - 1 - K^2)^2 + (\Omega^2/Q^2)(1 + \Psi^2\Omega^2 + 2\Psi K^2 Q)}.$$

Maximum output power is obtained when the EHs are shaken at the resonance frequency, and Equation (8.11) can be simplified as

$$P = \frac{\mu_1^2 ma^2}{2} \frac{K^2 Q^2}{(\omega_0 RC)^2 + (1 + \omega_0 RCK^2 Q)^2},$$

(8.14)

where μ_1 (= 1.566) is the correction factor for simple cantilever vibration at the first resonance [42]. The optimal load resistance R_{opt} is obtained from the differentiation of Equation (8.14) by R, and the maximum output power P_{opt} is obtained at R_{opt}; they are expressed as

$$R_{opt} = \frac{1}{\omega_0 C} \sqrt{\frac{1}{1+Q^2 K^4}} \qquad (8.15)$$

$$P_{opt} = \frac{\mu_1^2 m a^2}{2\omega_0} \frac{K^2 Q^2 \left(1+Q^2 K^4\right)}{1+\left(\sqrt{1+Q^2 K^4} + K^2 Q\right)^2}. \qquad (8.16)$$

This equation implies that, if mechanical and electrical values are given, output power can be determined by K and Q. The GEMC factor, an important parameter for determining the output power of piezoelectric energy harvesters, is obtained from

$$K^2 = \frac{\omega_{short}^2 - \omega_{open}^2}{\omega_{short}^2}, \qquad (8.17)$$

where ω_{short} and ω_{open} are the resonance angular frequencies of the associated short and open circuits, respectively.

The energy harvesters were plotted in Figure 8.11 and calculated the output power for the KNN/Si and PZT/Si energy harvesters in the case of $K^2 = 1.7 \times 10^{-3}$ and 0.65×10^{-3}, respectively. The maximum output power and K^2 are higher for KNN/Si than for PZT/Si. The performance of KNN thin-film EHs is comparable to or better than that of PZT thin-film EHs. The acceleration dependence of the output power was measured with accelerations ranging from 1 to 20 m/s^2 with the optimal load resistance, and the results are shown in Figure 8.13. For both EHs, the output voltage proportionally increases, while output power increases with the square of the acceleration, respectively. These results almost follow the theoretical prediction. The KNN and PZT EHs generated stable electric power under the high acceleration to 20 m/s^2 and it reached 3.7 and 3.4 µW at 20 m/s^2, respectively.

Because we used simple unimorph structures to evaluate the fundamental characteristics of the films, our observed output power and GEMC factors are smaller than those of other piezoelectric MEMS energy harvesters reported to date. However, the output power of both types of EHs can easily be enhanced by attaching a seismic mass to the cantilever tip and by further optimizing the harvester design and structure.

8.4.3 Metal-Based Piezoelectric MEMS Energy Harvesters

Piezoelectric MEMS EHs of metal cantilevers are advantageous to the long-term stability of the power generation due to excellent toughness of the metal. Furthermore, strong toughness of the metal cantilevers enables the reduction of the thickness of the cantilever and heavy tip mass attachment, which lead to the reduction of the resonant frequency and enhancement of vibration. In this study, PZT thin films were directly deposited on stainless steel cantilevers of ferritic stainless steel (SS430) cantilevers.

The coefficients of thermal expansion (CTE) for SS430 and PZT are about 11.4×10^{-6} K^{-1} and 9×10^{-6} K^{-1}, respectively. Considering that the CTE of typical austenitic stainless steel (SS304) is 18.4×10^{-6} K^{-1}, the small difference in CTE between PZT and SS430 is expected to mitigate the initial bending of the unimorph cantilevers. The fabrication process is shown in Figures 8.14 and 8.15. First, 30-μm thick SS430 cantilevers with tip masses were fabricated from a 300-μm thick stainless steel plate in a two-step, double-sided spray etching process using ferric chloride solution. This

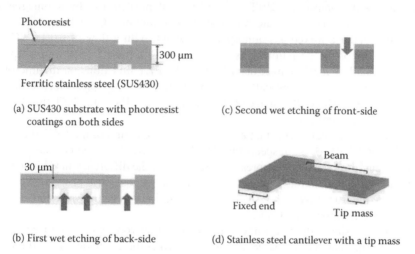

(a) SUS430 substrate with photoresist coatings on both sides

(b) First wet etching of back-side

(c) Second wet etching of front-side

(d) Stainless steel cantilever with a tip mass

FIGURE 8.14 Fabrication process of the stainless steel cantilevers with a tip mass by two-step spray etching.

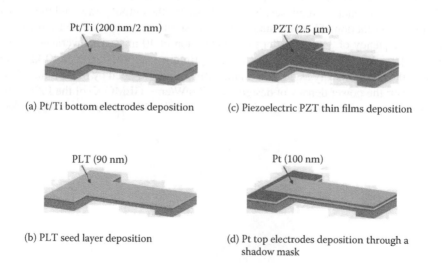

(a) Pt/Ti bottom electrodes deposition

(b) PLT seed layer deposition

(c) Piezoelectric PZT thin films deposition

(d) Pt top electrodes deposition through a shadow mask

FIGURE 8.15 Sputtering deposition of the piezoelectric PZT thin films on microfabricated stainless steel cantilever.

etching process enables fabrication of the fixed end and the tip mass of cantilevers without bonding or clamping. With the cantilever length fixed at 7.5 mm, cantilevers of different widths of 1.0, 2.0, and 5.0 mm were constructed, while the built-in tip mass varied from 5.0 to 25 mg.

Prior to PZT deposition, Pt/Ti bottom electrodes were deposited by RF-magnetron sputtering onto the microfabricated cantilevers. A (Pb, La)TiO$_3$ (PLT) seed layer was deposited to promote the growth of the PZT thin films, and successively the PZT thin films with composition Zr/Ti = 53/47 were deposited also by RF-magnetron sputtering [43,44]. The substrate was heated to around 600°C, and deposition was performed under a mixed gas atmosphere of Ar/O$_2$ with a flow of 9.0/1.0 SCCM. The thickness of the PZT films was 2.5 μm. The Pt top electrodes were prepared through a shadow mask. Direct deposition of the PZT thin films on the microfabricated cantilever could drastically simplify fabrication of the piezoelectric MEMS EHs. Figure 8.16 shows the image of the piezoelectric EHs of the PZT thin film on a stainless steel cantilever.

The relative dielectric constant ε_r and dielectric loss tan δ of the PZT films were 325–445 and 2.1%–3.3%, respectively. The low relative dielectric constant of PZT films is caused by in-plane compressive stress due to the difference in CTE between PZT thin films and stainless steel. The transverse piezoelectric properties of PZT films were measured from the deflection of the unimorph cantilevers, and $e_{31,f}$ was calculated to be about −4.0 C/m^2.

Figure 8.17 shows the frequency response of output voltage of a unimorph cantilever (7.5 mm long, 5.0 mm wide, 25 mg tip mass) at the acceleration of 10 m/s^2. A clear peak of the output voltage appeared at a resonance frequency of 367 Hz. The Q value was calculated to be 80 from the half-power bandwidth method. As shown in Figure 8.17, the nonlinear response (jump phenomenon and hysteresis) was clearly observed arising from softening spring effects; however, it is preferable in expanding the range of resonance frequency. Figure 8.18 shows the output power and the output voltage as a function of load resistance. The measurement was performed at the resonance frequency of 367 Hz under the acceleration of 10 m/s^2. The averaged output power [$P = V^2/2R$] reached 6.8 μW at the optimum load resistance of 10 kΩ. We define the power density as the maximum output power divided by the volume of the cantilever; the power density of design 1 is 1.7 mW/cm^3. GEMC K^2 of the PZT films on stainless steel cantilever was calculated to be 2.8 × 10^{-3} and almost independent

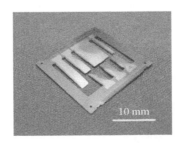

FIGURE 8.16 Photograph of piezoelectric EH of PZT thin film on microfabricated stainless steel cantilevers.

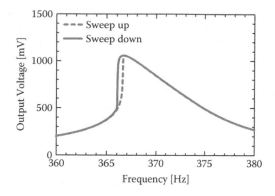

FIGURE 8.17   Frequency response of piezoelectric EH of PZT thin film on stainless steel cantilever.

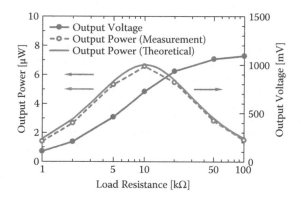

FIGURE 8.18   Output voltage and generated power as a function of load resistance.

of the design of the unimorph cantilevers. Compared with K^2 for PZT films on Si cantilevers (0.65×10^{-3}), we achieved an extremely high conversion efficiency for our PZT energy harvesters on stainless steel cantilevers.

8.5   SUMMARY

In this chapter, the outline of piezoelectric MEMS technologies is presented. Ferroelectric materials have a variety of functionality, including piezoelectricity, and integration of piezoelectric materials on MEMS enhances the function of the devices even in the simple microstructures. For the development of piezoelectric MEMS, deposition, evaluation, and microfabrication of piezoelectric thin films should be established. PZT is the most popular piezoelectric material, and PZT thin films with a thickness of a few microns are used for the MEMS applications. RF-magnetron sputtering is a suitable method for the deposition of piezoelectric PZT films. Transverse piezoelectric properties of the thin films are indispensable to design the piezoelectric MEMS, and they can be evaluated from the actuator properties of unimorph cantilevers of a PZT film and a substrate. On the other hand, lead-free piezoelectric thin films have been investigated as environmentally friendly materials, and it was reported that KNN thin films show large piezoelectric properties almost compatible with those of PZT films. Among a variety of piezoelectric MEMS applications, MEMS EHs have attracted attention for acting as an autonomous energy source for microdevices, especially for wireless sensor nodes. For the practical application, robust, environmentally friendly, low-cost, and high-efficiency MEMS EHs are required. Progress of the piezoelectric thin-film technologies will be a key to solving these issues.

REFERENCES

1. B. Piekarski, M. Dubey, E. Zakar, R. Polcawich, D. DeVoe, and D. Wickenden. 2002. Sol-gel PZT for MEMS applications. *Integrated Ferroelectrics* 42:25–37.
2. C. M. Foster, G. R. Bai, R. Csencsits, J. Vetrone, R. Jammy, L. A. Wills, E. Carr, and J. Amano. 1997. Single-crystal $Pb(Zr_x,Ti_{1-x})O_3$ thin films prepared by metal–organic chemical vapor deposition: Systematic compositional variation of electronic and optical properties. *Journal of Applied Physics* 81:2349–2357.
3. I. Vrejoiu, G. Le Rhun, L. Pintilie, D. Hesse, M. Alexe, and U. Gösele. 2006. Intrinsic ferroelectric properties of strained tetragonal $PbZr_{0.2}Ti_{0.8}O_3$ obtained on layer-by-layer grown, defect-free single-crystalline films. *Advanced Materials* 18:1657–1661.
4. I. Kanno, S. Fujii, T. Kamada, and R. Takayama. 1997. Piezoelectric properties of c-axis oriented $Pb(Zr,Ti)O_3$ thin films. *Applied Physics Letters* 70:1378–1380.
5. F. Calame, and P. Muralt. 2007. Growth and properties of gradient free sol-gel lead zirconate titanate thin films. *Applied Physics Letters* 90:062907.
6. T. Matsunaga, T. Hosokawa, Y. Umetani, R. Takayama, and I. Kanno. 2002. Structural investigation of $Pb_y(Zr_{0.57}Ti_{0.43})_{2-y}O_3$ films deposited on Pt(001)/MgO(001) substrates by RF sputtering. *Physics Review B* 66:064102.
7. I. Kanno, H. Kotera, T. Matsunaga, and K. Wasa. 2005. Intrinsic crystalline structure of epitaxial $Pb(Zr,Ti)O_3$ thin films, *Journal of Applied Physics* 97:074101.
8. J. F. Shepard, Jr., F. Chu, I. Kanno, and S. T. McKinstry. 1999. Characterization and aging response of the d_{31} piezoelectric coefficient of lead zirconate titanate thin films. *Journal of Applied Physics* 85:6711–6716.
9. E. Fujii et al. 2007. Preparation of (001)-oriented $Pb(Zr,Ti)O_3$ thin films and their piezoelectric applications. *IEEE Transactions on Ultrasonics Ferroelectrics Frequency Control* 54:2431–2438.
10. I. Kanno, S. Hayashi, T. Kamada, M. Kitagawa, and T. Hirao. 1993. Low-temperature preparation of $Pb(Zr,Ti)O_3$ thin films on $(Pb,La)TiO_3$ buffer layer by multi-ion-beam sputtering. *Japanese Journal of Applied Physics* 32:4057–4060.
11. T. Morimoto, O. Hidaka, K. Yamakawa, O. Arisumi, H. Kanaya, T. Iwamoto, Y. Kumura, I. Kunishima, and S. Tanaka. 2000. Ferroelectric properties of $Pb(Zi, Ti)O_3$ capacitor with thin $SrRuO_3$ films within both electrodes. *Japanese Journal of Applied Physics* 39:2110–2113.
12. K. Tomioka, F. Kurokawa, R. Yokokawa, H. Kotera, K. Adachi, I. Kanno. 2012. Composition dependence of piezoelectric properties of $pb(zr,ti)o_3$ films prepared by combinatorial sputtering, *Japanese Journal of Applied Physics* 51:09LA12.
13. S. H. Baek et al. 2011. Giant piezoelectricity on Si for hyperactive MEMS. *Science* 334:958.
14. J.-M. Liu, B. Pan, H. L. W. Chan, S. N. Zhu, Y. Y. Zhu, and Z. G. Liu. 2002. Piezoelectric coefficient measurement of piezoelectric thin films: An overview. *Materials Chemistry and Physics* 75:12–18.

15. G. Zavala, J. H. Fendler, and S. T. McKinstry. 1997. Characterization of ferroelectric lead zirconate titanate films by scanning force microscopy. *Journal of Applied Physics* 81:7480–7491.

16. A. L. Kholkin, C. Wutchrich, D. V. Taylor, and N. Setter. 1996. Interferometric measurements of electric field-induced displacements in piezoelectric thin films. *Review of Scientific Instruments* 67:1935–1941.

17. J. G. Smits, and W. Choi. 1991. The constituent equations of piezoelectric heterogeneous bimorphs. *IEEE Transactions on Ultrasonics Ferroelectronics Frequency Control* 38:256–270.

18. M. A. Dubois, and P. Muralt. 1999. Measurement of the effective transverse piezoelectric coefficient $e_{31,f}$ of AlN and $Pb(Zr_x,Ti_{1-x})O_3$ thin films. *Sensors and Actuators A* 77:106–112.

19. I. Kanno, H. Kotera, and K. Wasa. 2003. Measurement of transverse piezoelectric properties of PZT thin films. *Sensors and Actuators A* 107:68–74.

20. D.-M. Chun, M. Sato, and I. Kanno. 2013. Precise measurement of the transverse piezoelectric coefficient for thin films on anisotropic substrate. *Journal of Applied Physics* 113:044111.

21. T. Takenaka, H. Nagata, and Y. Hiruma. 2007. Current developments and prospective of lead-free piezoelectric ceramics. *Japanese Journal of Applied Physics* 47:3787–3801.

22. P. K. Panda. 2009. Review: Environmental friendly lead-free piezoelectric materials. *Journal of Materials Science* 44:5049–5062.

23. T. Takenaka, K. Maruyama, and K. Sakata. 1991. $(Bi_{1/2}Na_{1/2})TiO_3$-$BaTiO_3$ system for lead-free piezoelectric ceramics. *Japanese Journal of Applied Physics* 30:2236–2239.

24. L. Egerton, and D. M. Dillon. 1959. Piezoelectric and dielectric properties of ceramics in the system potassium-sodium niobate. *Journal of American Ceramics Society* 42:438–442.

25. W. Liu, and X. Ren. 2009. Large piezoelectric effect in Pb-free ceramics. *Physics Review Letters* 103:257602.

26. Y. Saito, H. Takao, T. Tani, T. Nonoyama, K. Takatori, T. Homma, T. Nagaya, and M. Nakamura. 2004. Lead-free piezoceramics. *Nature* 432:84–87.

27. T. Mino, S. Kuwajima, T. Suzuki, I. Kanno, H. Kotera, and K. Wasa. 2007. Piezoelectric properties of epitaxial $NaNbO_3$ thin films deposited on $(001)SrRuO_3$/Pt/MgO substrates. *Japanese Journal of Applied Physics* 46:6960–6963.

28. I. Kanno, T. Mino, S. Kuwajima, T. Suzuki, H. Kotera, and K. Wasa. 2007. Piezoelectric properties of $(K,Na)NbO_3$ thin films deposited on $(001)SrRuO_3$/Pt/MgO substrates. *IEEE Transactions on Ultrasonics, Ferroelectrics, Frequency Control* 54:2562–2566.

29. A. Onoe, A. Yoshida, and K. Chikuma. 1996. Heteroepitaxial growth of $KNbO_3$ single-crystal films on $SrTiO_3$ by metalorganic chemical vapor deposition. *Applied Physics Letters* 69:167–169.

30. S. I. Khartsev, M. A. Grishin, and A. M. Grishin. 2005. Characterization of heteroepitaxial $Na_{0.5}K_{0.5}NbO_3$/$La_{0.5}Sr_{0.5}CoO_3$ electro-optical cell. *Applied Physics Letters* 86:062901.

31. T. Saito, T. Wada, H. Adachi, and I. Kanno. 2004. Pulsed laser deposition of high-quality $(K,Na)NbO_3$ thin films on $SrTiO_3$ substrate using high-density ceramic targets. *Japanese Journal of Applied Physics* 43:6627–6631.

32. K. Shibata, F. Oka, A. Ohishi, T. Mishima, and I. Kanno. 2008. Piezoelectric properties of $(K,Na)NbO_3$ films deposited by RF magnetron sputtering. *Applied Physics Express* 1:11501.

33. K. Shibata, F. Oka, A. Nomoto, T. Mishima, and I. Kanno. 2008. Crystalline structure of highly piezoelectric $(K,Na)NbO_3$ films deposited by RF magnetron sputtering. *Japanese Journal of Applied Physics* 47:8909–8913.

34. I. Kanno, Y. Tazawa, T. Suzuki, and H. Kotera. 2007. Piezoelectric unimorph miroactuators with X-shaped structure composed of PZT thin films. *Microsystems Technology* 13:825–829.
35. Y. Kokaze, I. Kimura, M. Endo, M. Ueda, S. Kikuchi, Y. Nishioka, and K. Suu. 2007. Dry etching process for Pb(Zr,Ti)O$_3$ thin-film actuators. *Japanese Journal of Applied Physics* 46:280–282.
36. F. Kurokawa, I. Kanno, R. Yokokawa, H. Kotera, F. Horikiri, K. Shibata, T. Mishima, and M. Sato. 2012. Microfabrication of lead-free (K,Na)NbO$_3$ piezoelectric thin films by dry etching. *Micro & Nano Letters* 113:1223–1225.
37. S. Roundy, and P. K. Wright. 2004. A piezoelectric vibration based generator for wireless electronics. *Smart Materials Structures* 13:1131–1142.
38. S. P. Beeby, M. J. Tudor, and N. M. White. 2006. Energy harvesting vibration sources for microsystems applications. *Measurement Science and Technology* 17:R175–R195.
39. S-G. Kim, S. Priya, and I. Kanno. 2012. Piezoelectric MEMS for energy harvesting. *MRS Bulletin* 37:1039–1050.
40. S. Roundy, P. K. Wright, and J. Rabaey. 2003. A study of low level vibrations as a power source for wireless sensor nodes. *Computer Communications* 26:1131–1144.
41. M. Renaud, K. Karakaya, T. Sterken, P. Fiorini, C. Van Hoof, and R. Puers. 2008. Fabrication, modeling and characterization of MEMS piezoelectric vibration harvesters. *Sensors and Actuators A* 145–146:380–386.
42. A. Erturk, and D. J. Inman. 2008. On mechanical modeling of cantilevered piezoelectric vibration energy harvesters. *Journal of Intelligent Material Systems and Structures* 19:1311–1325.
43. T. Suzuki, I. Kanno, J. J. Loverich, H. Kotera, and K. Wasa. 2006. Characterization of Pb(Zr, Ti)O$_3$ thin films deposited on stainless steel substrates by RF-magnetron sputtering for MEMS applications. *Sensors and Actuators A* 125:382–386.
44. K. Kanda, I. Kanno, H. Kotera, and K. Wasa. 2009. Simple fabrication of metal-based piezoelectric MEMS by direct deposition of Pb(Zr,Ti)O$_3$ thin-films on titanium substrates. *Journal of Microelectromechanical Systems* 18:610–615.

Piezoelectric Vibration
Energy Harvesters
*Modeling, Design, Limits,
and Benchmarking*

A. Dompierre, S. Vengallatore, and L. G. Fréchette

CONTENTS

Abstract: This chapter introduces a lumped modeling and design approach
for bending beam type resonant piezoelectric vibration energy harvesters. The
developed framework is used to clarify the achievable power density limits of
these devices. Four possible bottlenecks are discussed: the amount of energy
transferred to the harvester from the source (inertial coupling), the amount
of strain energy that the structure can support (stress), what fraction of this
energy is converted electrically (piezoelectric conversion), and the fraction
that is scavenged by the electric load (energy extraction). For optimal resonator designs, the characteristics of the vibration source (acceleration and
frequency), the density of the materials used, and the quality factor of the
device define the achievable power density, suggesting 10–100 μW/cm^3 for
typical vibrations. The materials and configuration of the beam, stress and
fatigue considerations, and the circuitry then define the necessary piezoelectric material volume required to provide this power level. Figures of merit are
proposed to evaluate these characteristics independently of the operating conditions. Analyzing results from recently demonstrated resonant devices, the
inertial limit appears to be commonly achieved, but varying the beam design
and materials can significantly impact the beam power density. To close this
chapter, we briefly address the topic of wideband vibration energy harvesting.

9.1   INTRODUCTION

The development of novel portable sources of electrical power has been a focus of
scientific research for several years, in great part motivated by the progress and new
needs in the sectors of consumer electronics and wireless sensors. Recent developments in wireless communications and the miniaturization of electromechanical
devices using complementary metal–oxide semiconductor (CMOS) and microelectromechanical systems (MEMS) technologies are key enablers for very low
power electronics. Indeed, advances in these fields suggest that many intelligent
embedded devices will soon operate on very small power budgets, in the range of
1–100 μW [1]. Although electrochemical batteries are a mature technology that is
currently widely used, it restricts the performance and autonomy of microdevices in
many important emerging applications. For instance, it is not a practical option for
large-scale wireless sensor networks. In fact, the widespread deployment of these
sensor networks for health monitoring applications and the Internet of Things has
now hit a roadblock due to the cumbersome task of replacing so many batteries [2].

Hence, there is great interest in developing miniaturized energy harvesters that
can convert ambient energy—thermal fluctuations and gradients, solar radiation,

mechanical vibrations, fluid flow, and radio frequency waves—into electrical power. Over the past 20 years, numerous device concepts have been proposed, ranging from miniaturized engines and fuel cells to thermoelectric, thermophotovoltaic, and piezoelectric devices. Many crucial questions remain open in this field: What are the performance limits of different types of harvesters? Can we develop effective strategies for selecting harvesters for specific applications and environments? Can we develop rational design methodologies to optimize performance, reliability, and manufacturability? Can simple models be used to do this efficiently with good accuracy?

In this chapter, we seek answers to some of these questions for one particular type of harvester—namely, the piezoelectric vibration energy harvester (PVEH). The ubiquitous nature of vibrations makes this source a very interesting prospect for energy harvesting. For example, a VEH does not require direct access to ambient wind or sunlight sources, which can be rare in some environments. It also operates on the same principle used by accelerometers. Under the influence of an external vibration, a suspended mass deflects from its neutral position due to its own inertia. For a PVEH, the stresses induced in a piezoelectric material resulting from this deflection are used as the means of energy conversion. Much like their sensor counterpart, MEMS harvesters generally use cantilever or clamped-clamped beam geometries on which a large proof mass is attached. This mass fills two main functions: tune the natural frequency of the structure and increase the applied force and therefore the power output. Thin film deposition processes are often used for the integration of the piezoelectric material into these small devices.

Several high- and low-order models for PVEHs have been proposed through the years, but conflicting notions have spread in the literature [3]. Thankfully, recent models accurately describe the piezoelectric effect [4,5], but they are expressed in such ways that they lose the simplicity and transparency provided by lower order models such as the model of Williams and Yates [6]. A model with these attributes is nonetheless a valuable tool to streamline the design and optimization of harvesters. Moreover, many research groups in academic and industrial research laboratories have designed and built miniaturized PVEH devices using different configurations, geometries, materials, and electrical circuits. Due to their different design, their performances are tested under different operating conditions, making comparison among them difficult. Still, typical power densities of the order of microwatts per cubic centimeter have been reported, which is significantly lower than the early estimates that motivated efforts to build PVEHs [7,8]. Although increasing their power density is a major focus of current research, it remains unclear whether there is indeed any potential for significant improvements. Recognizing the limits of VEHs is important to better evaluate which wireless sensor applications are compatible with this technology.

The first objective of this book chapter is therefore to introduce a simple but accurate model for the PVEH. We subsequently present a global framework for design and analysis by exploring in each section the various aspects that can limit this type of energy harvesting. The model is used to quantify their respective impacts, with a special focus on the notion of power density (i.e., power per unit volume), a metric typically used to report the performances of devices. Moreover, these limits are used to propose relevant figures of merit (FOMs) for benchmarking and comparison of harvesters with different designs. The selection of an optimal piezoelectric material and electrical interface is also among the topics of this discussion. Before introducing this framework, we first take a look at the energy flow in PVEH devices.

9.1.1 Energy Transfer in PVEH Devices

Figure 9.1 presents a diagram of the energy flow for a general piezoelectric energy harvester application. The sequence starts from the ambient mechanical energy source, which interacts with the harvesting device. By doing work on the harvester, either by an inertial force (e.g., vibration, shocks) or a contact force (e.g., impacts, fluid flows, pressure gradients), the source transfers some of its energy to the device. Part of the transferred energy is present as kinetic energy (from the mass motion) and potential energy (from the elastic energy in the strained spring and the electrical energy in the charged capacitance), with an exchange between these two forms as the harvester vibrates. Due to the piezoelectric effect, an electrical potential is generated on the electrodes of the piezoelectric element while it is stressed, enabling conversion of some of the mechanical energy into electrical energy. Losses occur during this process (e.g., mechanical damping and dielectric leakage), while electrical energy is also extracted to be consumed, processed, or stored. If electrical energy remains, it is restored mechanically as kinetic energy. The extracted electrical energy can finally

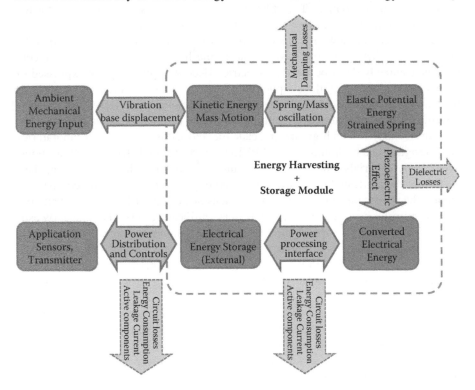

FIGURE 9.1   General energy chain of a PVEH device.

be used by an electronic device after appropriate processing by the power distribution and control interfaces.

This chain contains bidirectional arrows to show that the flow of energy can go in both ways. For instance, the control interface could inject previously stored energy into the piezoelectric element to tune the harvester electrically [9]. This tuning could therefore affect the capture of ambient energy by the harvester. The harvester also exchanges energy with its environment, but for the miniaturized devices under consideration, its effects on the source are assumed negligible. Looking at this chain, several limiting aspects can be identified as potential bottlenecks on the net electric power output and the power density:

- How much energy does the harvester capture? The vibration source and the quality of its mechanical coupling with the device have a tremendous impact on the harvestable power. Is the device excited by inertia or contact forces? Is transfer of energy based on resonance, impacts, or plucking? This limit is specifically discussed for linear resonators in Section 9.3.
- What is the maximum amount of energy that a device can collect without risks of damages? Stress concerns (i.e., distribution and concentration), fatigue, and materials degradation limit how much energy they can store before failure. This aspect, in theory, puts the upper bound on the power density, assuming that plenty of energy can be captured by the harvesting device in the first place. These issues are discussed in Section 9.4.
- How much mechanical energy can be converted into electrical energy? The generalized effective piezoelectric coupling factor, k_e^2, is a design parameter of piezoelectric transducers that characterizes the fraction of the applied energy that is transformed electrically under a specific strain distribution. As we discuss in Section 9.5, this coupling factor depends on material properties—mainly the material coupling factor expressed by its k_{ij} constants and the device geometry.
- How much of this electrical energy can be extracted and stored? Even if a large part of the work applied on the device is converted into electrical energy, only a fraction might be extractable. The load connected to the transducer plays a big role in this process. Many harvesting interfaces have been proposed to improve energy extraction and reduce losses, and their impact is presented in Section 9.6.

In addition, the design of the device dictates how these aspects are coupled to each other. For instance, typical inertial, resonant PVEHs are usually made of a suspended structure, often composed of multiple layers, which is clamped to the vibration source. Due to the inertia of the mass attached to this structure, stresses are induced to the materials. By designing this structure to match one of its resonant modes with the vibration source frequency, more energy can be captured, in turn inducing larger stresses as well. Hence, these influences are particularly strong here because a single heterogeneous structure captures, receives, and converts the energy. Moreover, energy is captured and extracted at the same time.

In Section 9.7, we use multiple FOMs to compare the performance of various devices reported in the literature and paint a picture of the current state of the art. Other considerations and perspectives, such as novel design approaches and trade-offs, are finally discussed in Section 9.8 to close this chapter.

9.2 SINGLE DEGREE OF FREEDOM MODEL OF A PVEH

Piezoelectric resonant structures can be modeled accurately using distributed parameter methods [4], but a simpler, lumped element approach is often sufficient and more convenient for design around a single resonant mode. Such a model will be first presented here to introduce the dynamics, followed by a more expansive analysis of the parameters' effects on the power output.

9.2.1 LUMPED PARAMETER MODELING

The lumped representation of the piezoelectric resonator is similar to the more traditional mechanical resonator. It partly consists of a mechanical spring, K_m; an equivalent mass, M_{eq}; and a mechanical damper, C_m. However, an electrically coupled spring, K_{el}, and an electrical damper, C_{el}, are now introduced to wholly capture the piezoelectric effect [10]. Tied to the piezoelectric coupling, these two elements are also affected by the electrical load that is connected to the piezoelectric element, with a specific frequency response due to the capacitive nature of the piezoelectric element. Figure 9.2 illustrates the lumped modeling approach, shown here for a piezoelectric unimorph cantilever beam with a tip mass, M_t, connected to a resistive load, R_{eq} [11].

For this analysis, we shall consider the transverse vibration mode of a thin beam of length L and width b. We also neglect any rotary inertia effect from the beam or the tip mass. The thicknesses of the piezoelectric and support layers are respectively

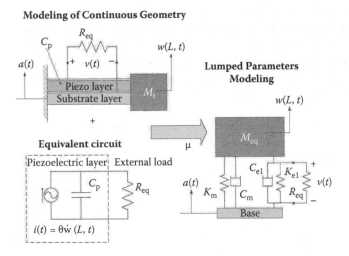

FIGURE 9.2 Equivalent lumped parameter model of a PVEH beam.

denoted by h_p and h_s, with their density and modulus of elasticity similarly denoted by ρ_p, ρ_s and Y_p, Y_s respectively. It is worth mentioning that, for thin beams, the elasticity is derived from the compliance of the material, such as [12]

$$Y = (s_{11}^E)^{-1},\qquad(9.1)$$

where $s_{11}^{E\,*}$ is the first term of the compliance tensor of the material. Let us further assume that the electrodes cover a length $L_1 \leq L$, starting from the base of the cantilever beam because this region will be the most stressed. Nevertheless, the electrodes are assumed to be much thinner than the beam; hence, their effects on stiffness and mass are negligible. The beam displacement is characterized by

$$w(x,t) = W\hat{w}(x)\sin(\omega t),\qquad(9.2)$$

where W is the tip displacement phasor (capturing the amplitude and phase information) and $\hat{w}$ is the function providing the shape of the deformed beam. This function is normalized to have $|\hat{w}(L)| = 1$. The energy source is a harmonic acceleration applied to the base of the cantilever beam:

$$a(t) = A\sin(\omega t),\qquad(9.3)$$

where A again expresses a phasor. The mechanical domain equation is based on Newton's second law and can be reduced to Equation (9.4). The electrical behavior is captured by Equation (9.5) and obtained from Kirchhoff circuit laws for an equivalent parallel resistor-capacitor (RC) circuit connected with a variable current source. Hence, the piezoelectric layer is equivalent to a strain rate-dependant current source and a capacitor, C_p, in the electrical domain.

$$M_{eq}\ddot{w}(L,t) + C_m\dot{w}(L,t) + K_m w(L,t) - \theta v(t) = f_{in}(t)\qquad(9.4)$$

$$\theta\dot{w}(L,t) + C_p\dot{v}(t) + \frac{v(t)}{R_{eq}} = 0\qquad(9.5)$$

The piezoelectric effect is therefore completely captured through the coupling force, θv, and piezoelectric current, $\theta\dot{w}(L,t)$, where $v(t)$ is the voltage and θ is the beam coupling coefficient. The equivalent lumped parameters are taken from Erturk and Inman [4] and Dompierre [13] and can be expressed as

$$\theta = Y_p d_{31}\left(b\bar{h}_p \int_0^{L_1} \frac{d^2\hat{w}(x)}{dx^2}dx\right),\qquad(9.6)$$

---

[*] The superscript E indicates that the property is measured with a constant or null electric field (i.e., in short-circuit condition).

$$C_p = \varepsilon_{33}^S \left(\frac{bL_1}{h_p} \right), \tag{9.7}$$

$$M_{eq} = M_t \hat{w}(L) + b \left(\rho_s h_s + \rho_p h_p \right) \int_0^L \left(\hat{w}(x) \right)^2 dx \tag{9.8}$$

$$K_m = \frac{b}{3} \left[Y_s (\overline{h}_b^3 - \overline{h}_a^3) + Y_p (\overline{h}_c^3 - \overline{h}_b^3) \right] \int_0^L \left(\frac{d^2 \hat{w}(x)}{dx^2} \right)^2 dx. \tag{9.9}$$

with

$$\overline{h}_a = -\frac{\dfrac{h_s}{2} Y_s h_s + \left(h_s + \dfrac{h_p}{2} \right) Y_p h_p}{Y_s h_s + Y_p h_p} \tag{9.10}$$

$$\overline{h}_b = h_s + \overline{h}_a \tag{9.11}$$

$$\overline{h}_c = h_p + \overline{h}_b \tag{9.12}$$

$$\overline{h}_p = \frac{\overline{h}_b + \overline{h}_c}{2}. \tag{9.13}$$

In these expressions, the terms $\overline{h}_a$, $\overline{h}_b$, and $\overline{h}_c$ are respectively the distance between the lower surface, the piezo/substrate interface, and the top surface from the neutral axis; $\overline{h}_p$ is the distance between the neutral axis and the midplane of the piezoelectric layer (see Figure 9.3). The 31-mode piezoelectric constant is denoted by d_{31} and the permittivity of the clamped capacitive piezoelectric layer is denoted by ε_{33}^S.[*]

As we can see, θ is proportional to d_{31} while both θ and K_m depend on geometry and strain distribution. This is an important point that is addressed in more detail in Sections 9.4 and 9.5. We now complete the introduction of the model by describing the inertial forcing term, $f_{in}(t)$, in the right-hand portion of Equation (9.4) [13]:

$$f_{in}(t) = -a(t) \left[M_t \hat{w}(L) + b \left(\rho_s h_s + \rho_p h_p \right) \int_0^L \hat{w}(x) dx \right]. \tag{9.14}$$

For a punctual mass, the applied force would instead be expressed by $a(t)M_{eq}$, but Equation (9.14) accounts for the effect of the mass distribution of the continuous

---

[*] The superscript S indicates that the property is measured with a constant strain.

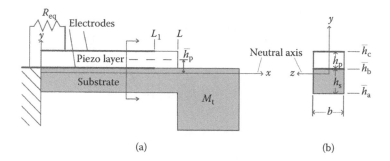

FIGURE 9.3 Considered beam geometry and relevant dimensions: (a) side view, and (b) cross section.

beam and the tip mass. Hence, to express the force as a function of the equivalent mass, we use a correction factor, noted μ, such that

$$f_{in}(t) = -M_{eq}a(t)\mu,$$ (9.15)

with

$$\mu = \frac{M_t\hat{w}(L) + b\left(\rho_s h_s + \rho_p h_p\right)\displaystyle\int_0^L \hat{w}(x)dx}{M_t\hat{w}(L) + b\left(\rho_s h_s + \rho_p h_p\right)\displaystyle\int_0^L \left(\hat{w}(x)\right)^2 dx}.$$ (9.16)

For typical cantilever configurations, the correction factor ranges from 1 (large tip mass) to 1.5 (no tip mass). Moreover, the equivalent mass of the lumped parameter model is smaller than the sum of the actual mass of the beam and the tip mass, M_t [3].

We can then express the electrical damping and stiffness explicitly using Laplace transforms and the phasor notation to combine these relationships. First, let us isolate the voltage in Equation (9.5) to obtain

$$V = -\left(\frac{j\omega R_{eq}}{1 + j\omega R_{eq}C_p}\right)\theta W,$$ (9.17)

where we multiply both the numerator and denominator by the complex conjugate of the denominator, $(1 - j\omega R_{eq}C_p)$, to bring the complex term at the numerator:

$$V = -\left(\frac{j\omega R_{eq} + \omega^2 R_{eq}^2 C_p}{1 + \omega^2 R_{eq}^2 C_p^2}\right)\theta W.$$ (9.18)

By substitution of the voltage in the Laplace transform of Equation (9.4), we obtain

$$\left\{ -\omega^2 M_{eq} + j\omega \left[C_m + \theta^2 \left(\frac{R_{eq}}{1+\omega^2 R_{eq}^2 C_p^2} \right) \right] \right.$$

$$\left. + \left[K_m + \theta^2 \left(\frac{\omega^2 R_{eq}^2 C_p}{1+\omega^2 R_{eq}^2 C_p^2} \right) \right] \right\} W = F_{in},$$ (9.19)

or, alternatively,

$$\left[-\omega^2 M_{eq} + j\omega (C_m + C_{el}) + (K_m + K_{el}) \right] W = F_{in},$$ (9.20)

where [10]

$$K_{el} = \frac{(\omega R_{eq} C_p)^2}{1+(\omega R_{eq} C_p)^2} \left(\frac{\theta^2}{C_p} \right),$$ (9.21)

$$C_{el} = \frac{(R_{eq} C_p)}{1+(\omega R_{eq} C_p)^2} \left(\frac{\theta^2}{C_p} \right).$$ (9.22)

Equations (9.21) and (9.22) show how the electrical stiffness and damping are tied to the relative impedance difference of the external load, R_{eq}, with respect to the capacitive layer, $(\omega C_p)^{-1}$. We discuss this behavior in more detail in Subsection 9.2.2. The tip displacement phasor, W, and the coupling force phasor, θV, can now be expressed as

$$W = \frac{\mu M_{eq} A}{(K_m + K_{el} - \omega^2 M_{eq}) + j\omega (C_m + C_{el})}$$ (9.23)

$$\theta V = -(K_{el} + j\omega C_{el}) W .$$ (9.24)

The undamped natural frequency, traditionally given by $\omega_n = \sqrt{K_m / M_{eq}}$, does not consider the effect of electrical stiffness. Therefore, the system's true resonant frequency is instead expressed by

$$\omega_r = \sqrt{\frac{K_m + K_{el}}{M_{eq}}} .$$ (9.25)

For convenience, each parameter can be expressed in dimensionless form. As defined in Equations (9.27)–(9.29), Ω is the frequency ratio, ζ_m the mechanical damping factor, α the dimensionless time constant, and κ^2 the dimensionless coupling ratio. This last parameter gives the ratio of electrical to mechanical potential energy in open circuit. It is also related to the effective coupling factor, k_e^2, via Equation (9.30) [14]:

$$\Omega = \frac{\omega}{\omega_n},$$ (9.26)

$$\zeta_{\mathrm{m}} = \frac{C_{\mathrm{m}}}{2\omega_n M_{\mathrm{eq}}},$$ (9.27)

$$\alpha = \omega_n R_{\mathrm{eq}} C_{\mathrm{p}},$$ (9.28)

$$\kappa^2 = \frac{\theta^2}{K_{\mathrm{m}} C_{\mathrm{p}}} = \frac{\text{Electrical Energy}}{\text{Mechanical Energy}},$$ (9.29)

$$k_{\mathrm{e}}^2 = \frac{\kappa^2}{1+\kappa^2} = \frac{\text{Electrical Energy}}{\text{Applied Energy}}.$$ (9.30)

Particular attention should be given to avoid mixing up κ^2 and k_{e}^2. Indeed, they are sometimes confused because their value can be similar for devices with low piezoelectric coupling ($\kappa^2 \ll 1$), yet the discrepancy is notable when coupling is stronger. Finally, we rearrange Equations (9.24) and (9.23) to get partially dimensionless expressions for the output voltage and the tip deflection [11]:

$$V = -\frac{K_{\mathrm{m}}}{\theta}\left(\Delta\Omega_{\mathrm{el}}^2 + 2j\Omega\zeta_{\mathrm{el}}\right)W,$$ (9.31)

$$W = \frac{\mu A}{\omega_n^2\left[1+\Delta\Omega_{\mathrm{el}}^2 - \Omega^2 + 2j\Omega\left(\zeta_{\mathrm{m}}+\zeta_{\mathrm{el}}\right)\right]},$$ (9.32)

where $\Delta\Omega_{\mathrm{el}}^2$ and ζ_{el} are the dimensionless forms of the electrical stiffness and damping, respectively, given by

$$\Delta\Omega_{\mathrm{el}}^2 = \frac{K_{\mathrm{el}}}{K_{\mathrm{m}}} = \kappa^2\left(\frac{\Omega^2\alpha^2}{\Omega^2\alpha^2 + 1}\right),$$ (9.33)

$$\zeta_{\mathrm{el}} = \frac{C_{\mathrm{el}}}{2\omega_n M_{\mathrm{eq}}} = \frac{\kappa^2}{2}\left(\frac{\alpha}{\Omega^2\alpha^2 + 1}\right).$$ (9.34)

Physically, $\Delta\Omega_{\mathrm{el}}^2$ is responsible for the resonance frequency shift typically observed on piezoelectric devices. Hence, it can also be seen as a resonance frequency ratio shift from the natural frequency of the short-circuited device. A dimensionless form for Equation (9.25) is similarly obtained:

$$\Omega_{\mathrm{r}} = \omega_{\mathrm{r}}/\omega_n = \sqrt{1+\Delta\Omega_{\mathrm{el}}^2}.$$ (9.35)

The electrical power generated by the piezoelectric element can be evaluated by considering the equivalent impedance of the RC circuit shown in Figure 9.2

$$Z_{eq} = \left(\frac{1}{R_{eq}} + j\omega C_p \right)^{-1} = \frac{R_{eq}}{1 + j\omega R_{eq} C_p}. \tag{9.36}$$

The complex electric power, P_{el}, is therefore given by

$$P_{el} = \frac{VV^*}{2Z_{eq}^*} = \frac{|V|^2}{2Z_{eq}^*}, \tag{9.37}$$

where the asterisk denotes the complex conjugate. Based on Equations (9.31)–(9.34), we can expand and rearrange to get

$$
\begin{aligned}
P_{el} &= \frac{1}{2} \frac{K_m^2}{\theta^2} \left(\Delta\Omega_{el}^4 + 4\Omega^2\zeta_{el}^2 \right) \frac{\left(1 - j\omega R_{eq} C_p \right)}{R_{eq}} |W|^2 \\
&= \frac{1}{2} \frac{K_m^2 C_p}{\theta^2} \left[\kappa^4 \frac{\left(\Omega^2\alpha^2 \right)^2 + \Omega^2\alpha^2}{\left(\Omega^2\alpha^2 + 1 \right)^2} \right] \frac{\left(1 - j\omega R_{eq} C_p \right)}{R_{eq} C_p} |W|^2 \\
&= \frac{1}{2} \omega K_m \left(\kappa^2 \frac{\Omega^2\alpha^2}{\Omega^2\alpha^2 + 1} \right) \frac{\left(1 - j\Omega\alpha \right)}{\omega R_{eq} C_p} |W|^2 \\
&= \frac{1}{2} \omega K_m \kappa^2 \left(\frac{\Omega\alpha - j\Omega^2\alpha^2}{\Omega^2\alpha^2 + 1} \right) |W|^2.
\end{aligned}
\tag{9.38}
$$

The real part of this expression is called the active power, which is the actual average power dissipated by the resistance. In contrast, the complex part is reactive power—that is, electrical energy that is stored in the device but not extracted. Looking at Equations (9.33) and (9.34), it becomes clear that the active and reactive powers are respectively tied to the electrical damping and the electrical stiffness. Hence, the average harvestable power output, $\bar{P}_{el}$, can be rearranged as

$$
\begin{aligned}
\bar{P}_{el} &= \omega K_m \Omega \zeta_{el} |W|^2 \\
&= \frac{K_m \mu^2 |A|^2}{\omega_n^3} \frac{\Omega^2 \zeta_{el}}{\left[\left(1 + \Delta\Omega_{el}^2 - \Omega^2 \right)^2 + 4\Omega^2 \left(\zeta_m + \zeta_{el} \right)^2 \right]} \\
&= \frac{\mu^2 M_{eq} |A|^2}{\omega} \frac{\Omega^3 \zeta_{el}}{\left[(1 + \Delta\Omega_{el}^2 - \Omega^2)^2 + 4\Omega^2 \left(\zeta_m + \zeta_{el} \right)^2 \right]}.
\end{aligned}
\tag{9.39}
$$

The first term of Equation (9.39) regroups parameters such as the vibration amplitude and frequency, as well as the equivalent mass of the device. The second term provides the frequency response and the effect of damping on the output power.

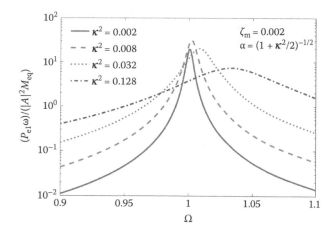

FIGURE 9.4 Effect of the coupling ratio, κ^2, on the power frequency response.

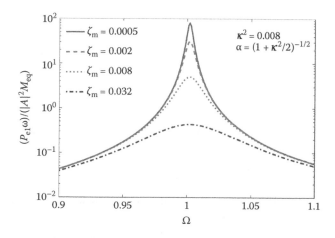

FIGURE 9.5 Effect of the mechanical damping factor, ζ_m, on the power frequency response.

Figures 9.4, 9.5, and 9.6 illustrate the frequency response trends for this second term with variations of κ^2, ζ_m, and α respectively. In the next subsection, we analyze further the reasons for the behavior observed.

9.2.2 PIEZOELECTRICITY EFFECTS ON RESONANCE AND CLOSED-LOOP SOLUTIONS

As previously discussed, the piezoelectric effect can be captured by the addition of a spring and a damper that are frequency and load dependent. The response of both components, provided by Equations (9.21) and (9.22), is tied to the relative difference of impedance between the harvester and the extracting load. Figure 9.7 illustrates the trends based on their dimensionless formulation. We observe that K_{el} increases with the relative impedance of the external load. This stiffening behavior occurs because the flow of electrical energy out of the electromechanical

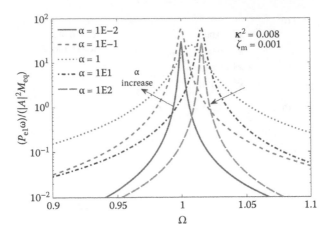

FIGURE 9.6   Effect of the electric load, α, on the power frequency response.

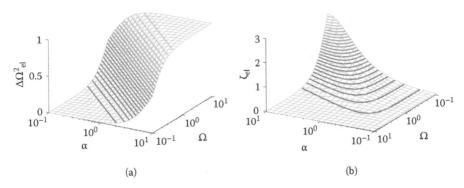

(a)   (b)

FIGURE 9.7   (a) Electrical stiffness and (b) damping trends with the load and frequency variations. The values are normalized by κ^2.

structure is progressively blocked, thereby increasing the stored potential energy in the material. Hence, the term θ^2/C_p effectively represents the maximum feedback piezoelectric force acting on the structure per displacement when the load is an open circuit. Therefore, it is possible to measure the effective coupling factor of the piezoelectric transducer directly by observing the frequency shift occurring between the short-circuit and open-circuit conditions. In contrast, the electrical damping is maximized when $R_{eq} = (\omega C_p)^{-1}$, which is the condition of impedance matching. We can express this in dimensionless form:

$$\alpha = \Omega^{-1}. \tag{9.40}$$

In addition, there is no electrical damping occurring in both the open- and short-circuit conditions.

Due to the simultaneous stiffening and dampening caused by the piezoelectric effect, the resonance frequency shifts with the increase of the impedance of the connected load. A closed-loop solution for the resonance frequency ratio that depends only on the load can be obtained by solving Equation (9.25) for $\Omega = \Omega_r$:

$$\Omega_r^2 = 1 + \frac{\Omega_r^2 \alpha^2 \kappa^2}{1 + \Omega_r^2 \alpha^2}$$

$$0 = \alpha^2 \Omega_r^4 + \Omega_r^2 \left[1 - \alpha^2 \left(1 + \kappa^2 \right) \right] - 1.$$

(9.41)

From the four solutions of Equation (9.41), two are complex and two are real. The only valid solution is the positive real root provided by

$$\Omega_r = \sqrt{\frac{\alpha^2 \left(1 + \kappa^2 \right) - 1 + \sqrt{\left[1 - \alpha^2 \left(1 + \kappa^2 \right) \right]^2 + 4\alpha^2}}{2\alpha^2}}.$$

(9.42)

For any Ω_r, an associated electrical damping factor (Equation 9.34) can also be expressed as a function of only the load:

$$\zeta_{el} \big|_{\Omega = \Omega_r} = \frac{\alpha \kappa^2}{1 + \alpha^2 \left(1 + \kappa^2 \right) + \sqrt{\left[1 - \alpha^2 \left(1 + \kappa^2 \right) \right]^2 + 4\alpha^2}}.$$

(9.43)

At the resonance, the expressions for the tip displacement and power output (Equations 9.32 and 9.39) can also be simplified to

$$|W|\big|_{\Omega = \Omega_r} = \frac{\mu |A|}{2\omega_n \omega_r \left(\zeta_m + \zeta_{el} \right)}$$

(9.44)

$$\bar{P}_{el} \big|_{\Omega = \Omega_r} = \frac{\mu^2 M_{eq} |A|^2}{4\omega_n} \frac{\zeta_{el}}{\left(\zeta_m + \zeta_{el} \right)^2}.$$

(9.45)

It is important to stress again that both expressions are implicitly dependent on only the load, through Equations (9.42) and (9.43). This expression can therefore be used effectively to find the optimal load.

9.2.3 RESISTIVE LOAD OPTIMIZATION

Equation (9.45) is optimized by solving

$$\frac{d |\bar{P}_{el}|\big|_{\Omega = \Omega_r}}{d\zeta_{el}} = 0$$

$$0 = \frac{\left(\zeta_m + \zeta_{el} \right)^2 - 2\zeta_{el} \left(\zeta_m + \zeta_{el} \right)}{\left(\zeta_m - \zeta_{el} \right)^4}.$$

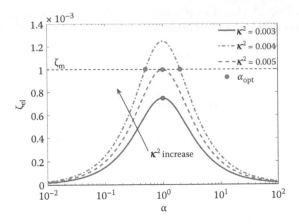

FIGURE 9.8 Electrical damping at resonance for several coupling ratios κ^2. The optimal loads are marked with circles and the dotted line is the corresponding mechanical damping level considered.

The result,

$$\zeta_{el} = \zeta_m, \tag{9.46}$$

shows that both damping components should be matched to optimize the power output—a conclusion that was previously demonstrated for electromagnetic harvesters [6]. Therefore, piezoelectric devices differ mainly by the significant stiffening behavior observed, which results in a shift of the resonance frequency. This result is also of significant importance to optimize the load connected to the PVEH device. Equation (9.43) is plotted on Figure 9.8 to observe the trend for the load-dependent electrical damping at resonance for several coupling ratios κ^2.

By comparison with a given mechanical damping level, shown in the figure by a dotted line, we observe that if the coupling ratio is small, matching both damping components is not possible due to the limited maximum electrical damping achievable. The power output is therefore maximized via a single load, which has a matched impedance to the piezoelectric layer. In contrast, a large coupling ratio leads to cases where the electrical damping is the main damping mechanism of the piezoelectric device. Two loads can now maximize the power output, a low-impedance load (close to the short-circuit condition, low-voltage output) and a high-impedance load (close to the open-circuit condition, high-voltage output). Between these two loads, the electrical damping dominates and we observe a reduction of the power output, but an increase of the device bandwidth (see Figure 9.6). Hence, beyond a certain coupling threshold, it is not possible to increase the power output for single-frequency harmonic sources. We later explain the reason for this phenomenon in Section 9.3.1.

The comparison of the coupling with the mechanical damping is, therefore, an effective method to identify the type of load optimization required. Thankfully, the piezoelectric resonator figure of merit can be used to accomplish this comparison.

Frequently used to characterize the performance of piezoelectric surface acoustic wave (SAW) devices, we reintroduce this FOM here in the context of VEHs.

9.2.4 RESONATOR FIGURE OF MERIT

The piezoelectric resonator FOM is expressed by $\kappa^2 Q_m$ [14], where Q_m is the mechanical quality factor of the resonator. The quality factor is more generally defined as

$$Q = \frac{1}{2\zeta} = \omega_r \frac{\text{Energy Stored}}{\text{Power Dissipated}}. \tag{9.47}$$

In an electromechanical system such as the piezoelectric resonator, the stored energy is the sum of the electrical energy stored in the capacitor, the potential energy stored in the spring, and the kinetic energy in the mass. The power dissipated is the sum of all the power dissipation sources. Since power can be dissipated mechanically or electrically, this definition is applicable to both mechanical damping and electrical damping. Hence, the total quality factor of the device can be expressed as

$$Q_{tot} = \left(\frac{1}{Q_m} + \frac{1}{Q_{el}} \right)^{-1}. \tag{9.48}$$

Using the equations introduced in this section, we can establish a coupling criterion to state where this transition occurs. Remember that the condition for which the maximum electrical damping occurs is when the impedances are matched (Equation (9.40)). Substitution of such a load in Equations (9.33) and (9.34) gives

$$\Delta\Omega_{el}^2 \big|_{\alpha=\Omega^{-1}} = \frac{\kappa^2}{2} \tag{9.49}$$

$$\zeta_{el} \big|_{\alpha=\Omega^{-1}} = \frac{\kappa^2}{4\Omega}. \tag{9.50}$$

Therefore, there is a unique frequency ratio and load for which there is both resonance and a maximization of the electrical damping:

$$\Omega_r \big|_{\alpha=\Omega^{-1}} = \sqrt{1 + \kappa^2/2}, \tag{9.51}$$

and the resulting maximum achievable electrical damping at resonance operation is expressed by

$$\zeta_{el} \big|_{\Omega=\Omega_r, \alpha=\Omega_r^{-1}} = \frac{\kappa^2}{4\sqrt{1 + \kappa^2/2}}. \tag{9.52}$$

When $\zeta_{el} \geq \zeta_m$, the transition occurs, which is to say that

$$\kappa^2 \geq 4\zeta_m \left(\zeta_m + \sqrt{\zeta_m^2 + 1} \right). \tag{9.53}$$

For low coupling systems, Equation (9.52) can be approximated to

$$\zeta_{el} \big|_{\Omega = \Omega_r, \alpha = \Omega_r^{-1}} \approx \frac{\kappa^2}{4}. \tag{9.54}$$

The error introduced by Equation (9.54) is less than 2.5% for $\kappa^2 < 0.1$. Equation (9.53) can also be reduced to

$$\kappa^2 > 4\zeta_m. \tag{9.55}$$

Finally, the value of the resonator FOM for which the transition occurs is approximately

$$\kappa^2 Q_m \approx 2. \tag{9.56}$$

As a criterion for PVEH, the resonator FOM basically states that devices with low piezoelectric coupling or a low mechanical quality factor ($\kappa^2 Q_m \ll 2$) may not be able to reach the optimal power generation condition. It is therefore useful to identify the design state of the resonating device. There is an optimal design state for which $\kappa^2 Q_m > 2$, where

- Mechanical damping is the only limiting factor.
- Mechanical damping and electrical damping should be matched by adjusting the load to maximize the power output at a single frequency resonance peak.
- There are two optimal loads and their values are strictly tied to mechanical damping and the coupling. These two values are obtained by solving Equation (9.43) for $\zeta_{el} = \zeta_m$ [13]:

$$\alpha_{opt} \approx \frac{1}{1+\kappa^2} \left\{ \frac{\kappa^2}{4\zeta_m} + \zeta_m \left[1 \pm \sqrt{1 - \frac{\kappa^2}{2\zeta_m^2} + \frac{1}{\zeta_m^2} \left[\left(\frac{\kappa^2}{4\zeta_m} \right)^2 - 1 \right]} \right] \right\}. \tag{9.57}$$

The second design state, for which $\kappa^2 Q_m \leq 2$, implies that

- Coupling and mechanical damping are both limiting factors.
- Electrical damping should be maximized.
- There is a single optimal load at resonance and it depends only on coupling:

$$\alpha_{opt} = \frac{1}{\sqrt{1 + \frac{\kappa^2}{2}}}. \tag{9.58}$$

With these equations in place, we are able fully to appreciate the relative importance of all the parameters in the optimization of the power output of resonant

TABLE 9.1

Baseline Parameters Used for Estimates

Young Modulus Y_p (GPa)	Dynamic Tensile Strength σ_t (MPa)	Coupling Factor k_{31}	Quality Factor Q_m	Frequency f (Hz)	Acceleration Amplitude $|A|$ (m/s²)	Density ρ_M (kg/m³)
81.3	24	0.33	250	150	1	2330

PVEHs and, more precisely, the influence of the coupling factor and the mechanical damping. We will now discuss the implications of these conclusions, in terms of fundamental limits and how they impact the device design, by considering geometry optimization, the choice of an interface for electrical energy extraction, and the selection of a suitable piezoelectric material. The equations developed will be used to give numerical estimates for those limits and the performance of a typical PVEH device. For future reference, the parameters found in Table 9.1 will be used for a baseline design. The piezoelectric material considered is a common hard lead zirconate titanate ceramic, PZT4 [15], and silicon is used as the tip mass because it can be readily micromachined. Also note that, in the generator mode, the piezoelectric material coupling factor, k_{31}, is the ratio between the electrical energy transformed to the energy provided (i.e., $k_{31}^2 = E_{el}/E_{applied}$). For a single-direction, 31-mode plane stress situation as in the previously introduced model [12,14]:

$$k_{31}^2 = \frac{d_{31}^2 Y_p}{\varepsilon_{33}^T},$$

(9.59)

$$\varepsilon_{33}^T = \varepsilon_{33}^S + d_{31}^2 Y_p,$$

(9.60)

where ε_{33}^T is the permittivity of the piezoelectric material measured with a constant and preferably null stress.

9.3 LIMIT BASED ON INERTIAL COUPLING

How much energy is transmitted to the device fundamentally restrains what it can harvest. In this section, we seek to evaluate how the intensity of the vibration source, the quality of the mechanical coupling and the process of harvesting energy limit this quantity for PVEH. First, let us assume that the piezoelectric device has a very large tip mass; hence, $M_{eq} \approx M_t$ and $\mu = 1$. The volume of this mass, Vol_M, occupies most of the space of the device and the effect of its sweeping motion is also neglected. We also assume that the frequency shift due to the piezoelectric effect is small; hence, $\omega_n \approx \omega_r$. Finally, we consider the situation of steady-state vibrations and therefore the transfer of energy into the harvester should be equal to what it dissipates:

$$\bar{P}_{in} = \bar{P}_{el} + \bar{P}_{m,loss}.$$

(9.61)

9.3.1 Inertial Coupling Limit of Linear Resonator

In its simplest form, the inertial vibration energy harvester is an electromechanical resonator excited by its own inertia. The actual power output expected from such a device is therefore fundamentally tied to the rate at which it can inertially capture energy from surrounding. To maintain the resonator steady-state oscillations, mechanical energy must be continuously supplied by the inertial force acting on the oscillating mass. The instantaneous power transfer is the result of the product of the applied force on the mass, $f_{in}(t)$, and its velocity, $\dot{w}(L,t)$. The average power transfer, $\bar{p}_{in}(t)$, is therefore given by

$$\bar{p}_{in} = \frac{1}{T}\int_0^T f_{in}(t)\dot{w}(L,t)dt, \tag{9.62}$$

where T is the observation time. In the frequency domain, the average complex vibration power input is obtained by computing the cross spectrum between these quantities [16,17]:

$$P_{in} = \frac{\dot{W}^* F_{in}}{2} = \frac{-j\omega W^* M_{eq} A}{2}. \tag{9.63}$$

After substitution of Equation (9.32) in Equation (9.63),

$$P_{in} = \frac{M_{eq}A}{2} \frac{-j\omega A^*}{\omega_n^2\left[1 + \Delta\Omega_{el}^2 - \Omega^2 - 2j\Omega\left(\zeta_m + \zeta_{el}\right)\right]}. \tag{9.64}$$

Equation (9.64) is then rearranged to bring all complex terms to the numerator:

$$P_{in} = \frac{M_{eq}|A|^2}{2\omega} \frac{\left[2\Omega^3\left(\zeta_m + \zeta_{el}\right) - j\Omega^2\left(1 + \Delta\Omega_{el}^2 - \Omega^2\right)\right]}{\left[\left(1 + \Delta\Omega_{el}^2 - \Omega^2\right)^2 + 4\Omega^2\left(\zeta_m + \zeta_{el}\right)^2\right]}. \tag{9.65}$$

Again, the real part of this complex power is the active power. It is the time average net power transferred to the resonator as described in Equation (9.66):

$$\bar{P}_{in} = \frac{M_{eq}|A|^2}{\omega} \frac{\Omega^3\left(\zeta_m + \zeta_{el}\right)}{\left[\left(1 + \Delta\Omega_{el}^2 - \Omega^2\right)^2 + 4\Omega^2\left(\zeta_m + \zeta_{el}\right)^2\right]}. \tag{9.66}$$

The imaginary part is the reactive power (i.e., energy temporarily stored in the resonator but later restored to the environment due to the phase difference between the inertial force and the mass displacement). At the resonance, when the applied force is perfectly in phase with the mass velocity, the active power is indeed maximized and the reactive power component is zero in Equation (9.65). After simplification for the case of resonance,

$$\bar{P}_{\text{in}}\Big|_{\Omega=\Omega_r} = \frac{M_{\text{eq}}|A|^2}{4\omega_n\left(\zeta_m + \zeta_{\text{el}}\right)}. \tag{9.67}$$

The presence of the electrical damping at the denominator of Equation (9.67) indicates that the power input is reduced as more energy is harvested. Hence, there is a backward effect, which explains the result observed in Section 9.2.3. If we consider matching of the damping components to optimize the power output, after substitution of the mechanical quality factor via Equation (9.47), Equation (9.67) becomes

$$\bar{P}_{\text{in}} = \frac{M_{\text{eq}}|A|^2 Q_m}{4\omega_n}. \tag{9.68}$$

In terms of mass volume, Vol_M,

$$\frac{\bar{P}_{\text{in}}}{Vol_M} = \frac{\rho_M|A|^2 Q_m}{4\omega_n}. \tag{9.69}$$

For the baseline parameters, this input mechanical power density is 154 µW/cm³. Let us now compare with the electrical power output (Equation (9.45)) by similarly expressing the electrical power density as

$$\frac{\bar{P}_{\text{el}}}{Vol_M} = \frac{\rho_M|A|^2}{4\omega_n}\frac{\zeta_{\text{el}}}{\left(\zeta_m + \zeta_{\text{el}}\right)^2}. \tag{9.70}$$

Again, for $\zeta_{\text{el}} = \zeta_m$, the theoretical electrical power density limit is reached [18]:

$$\frac{P_{\text{lim}}}{Vol_M} = \frac{\rho_M|A|^2 Q_m}{8\omega_n}. \tag{9.71}$$

Unsurprisingly, the resulting electrical power density is estimated to be 77 µW/cm³— exactly half the input mechanical power density.

This ratio of output electrical power to input mechanical power is the harvesting efficiency of the resonator and its expression is more generally obtained for any frequency from the ratio of Equation (9.39) on Equation (9.66):

$$\eta = \frac{\bar{P}_{\text{el}}}{\bar{P}_{\text{in}}} = \frac{\zeta_{\text{el}}}{\left(\zeta_m + \zeta_{\text{el}}\right)}. \tag{9.72}$$

The 50% efficiency obtained previously therefore means that half of the energy pumped into the resonator is lost through mechanical damping and the other half is extracted as electrical energy. In essence, if ζ_{el} is higher than ζ_m, a larger fraction of the energy pumped into the system and converted electrically will be extracted.

Moreover, the resonator FOM also relates to the maximum harvesting efficiency of the resonator. This is shown by simply replacing ζ_{el} by Equation (9.54) and ζ_m by its equivalent Q_m. With an impedance matched resistance, this efficiency is given by [19,20]

$$\eta_{max} \approx \frac{\kappa^2 Q_m}{2 + \kappa^2 Q_m}. \tag{9.73}$$

However, we have shown that increasing the efficiency over the 50% mark does not necessarily improve the electrical power. In fact, the excess damping reduces the amplitude, W, and the input mechanical power is lower. The electrical power output is then reduced due to this backward coupling effect we previously described. The device is more *efficient*, but it produces less power from the same source. In other words, it is also less *effective* at capturing energy and producing electrical power from a single-frequency harmonic source [21]. Therefore, another FOM, ξ, is defined to describe the harvesting effectiveness:

$$\xi = \bar{P}_{el}/P_{lim} = 4 \frac{\zeta_{el}}{\left(\zeta_m + \zeta_{el}\right)^2}. \tag{9.74}$$

Figure 9.9 shows the general trend expected for these FOMs as a function of the piezoelectric resonator FOM in the condition of matched impedances and resonance. For $\kappa^2 Q_m < 2$, both of the FOMs increase. However, once the coupling threshold is achieved, the resonant device can be operated to optimize either the efficiency (by matching the load impedance to the transducer) or the effectiveness (by adjusting the load in order to match the electrical damping with the mechanical damping) [21].

Let us consider a scenario where the energy dissipation is entirely caused by the piezoelectric damping, now twice as important as before (i.e., $\zeta_m = 0$ and $\zeta_{el} = 0.004$). The average mechanical power input would remain the same, because Q_{tot} has not

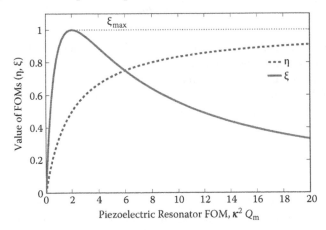

FIGURE 9.9 Variation of the FOMs with the resonator FOM, $\kappa^2 Q_m$.

changed, but the efficiency would jump to 100%. The harvested power would also be twice of what it was previously. Due to the energy dissipation being completely controlled by the piezoelectric damping, there would be theoretically no limit to $\bar{P}_{in}$ or P_{lim} and the effectiveness would now be zero. However, this result is not realistic because it requires an infinite mass displacement, and the harvester would eventually break due to the intense stresses. Hence, each figure of merit has its limit.

9.3.2 SUMMARY

Wrapping up what we have seen so far, we know that the energy involved in the operation of a resonator can be grouped in three categories:

1. The energy that flows into the device, from the inertial force that applies work on the oscillating mass. Resonance optimizes the capture of this energy because the applied force is in phase with the tip mass velocity.
2. The energy that flows out of the device, via various dissipation mechanisms such as mechanical damping losses, dielectric losses, and energy harvesting. In the model, the electrical power output is directly given by the electrical damping, but dielectric losses would also introduce additional electrical damping.
3. The energy that remains inside the harvester flows back and forth as potential and kinetic energy. During steady-state vibrations, the electrical power output will increase as long as the effect of the electrical damping is less than or equal to the effect of the mechanical damping. If $\zeta_{el} > \zeta_m$, the device is more efficient but the power output actually decreases because less energy is captured due to a backward coupling effect.

Figures 9.10 and 9.11, respectively, plot the evolution of those different energy and power categories in the time domain for three different regimes: (a) ramp up; (b) steady state, with simultaneous energy injection and extraction; and then (c) decay, with extraction only. A sinusoidal base acceleration with an amplitude of 1 m/s^2 is applied to the base of the system during the first 4 seconds. Once the excitation is stopped completely, no more energy is added and the stored energy is gradually dissipated or extracted. At this point, energy extraction would be optimized by maximizing electrical damping so that most of the energy remaining in the device can be harvested.

According to Equation (9.71), high values of Q_m and $|A|^2/\omega$ always lead to devices with high power densities, because more energy can be injected in both cases for the same mass volume. However, this result does not account for several factors. Among these is the increase of the necessary traveling distance of the mass. Indeed, if it is in the same range as the dimensions of the device, the estimation approach we used here becomes flawed and the displacement of the mass becomes an important concern in the design of the package [18].

Another point to consider is that more energy must be stored in high Q resonators, which implies important stresses. The design of the strained piezoelectric structure has not been considered at all so far, but there is a limit on the amount of strain that it can support and to the amount of energy that it can convert. These additional limits on performance are the focus of the next sections.

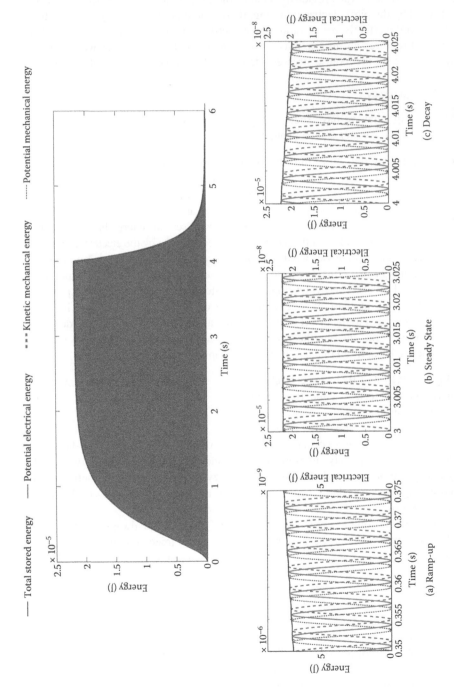

FIGURE 9.10 Evolution in the time domain of the different types of energy for different regimes: $\zeta_m = 0.002$, $\zeta_{el} = 0.0005$.

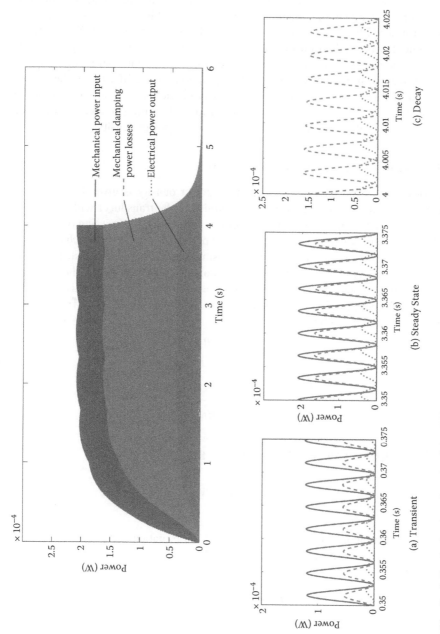

FIGURE 9.11 Evolution in the time domain of the different types of power for different regimes. $\zeta_{sm} = 0.002$, $\zeta_{sel} = 0.0005$.

9.4   STRESS-BASED LIMITS

The ultimate tensile stress of piezoceramics can be as high as a few hundred mega-pascals. In fact, this value can increase further for small-scale structures because they can be prepared with higher crystalline quality and lower defect densities than bulk ceramics [22]. However, for some materials, only a fraction (20% to 30%) of this value can be applied during operation to avoid degradation or failure due to depolarization and fatigue [23,24]. For this reason, the maximum stress in the spring element must be limited to ensure reliable operation for the lifetime of the device. The goal of this section is to quantify the effect of such a constraint on the power density. We first examine the best-case scenario of the uniform stress distribution and then observe the effect of a stress gradient on the power density.

9.4.1   POWER DENSITY WITH UNIFORM STRESS

We consider that the maximum mechanical power density is given by the rate at which the maximum strain energy can be applied to the transducer without leading to its failure. Hence, the optimal power density is achieved by stressing the material to this limit, here noted by σ_l, as uniformly as possible. To represent such a uniform stress field, we first assume that the device is simply made of a piezoelectric block of length L, thickness h, and width b, stressed along its axis. It is clamped on one end and pressure is applied to its other end. The stress amplitude is σ_l and it is applied in a sinusoid fashion, cycling between tension and compression. The stiffness of this block is therefore given by

$$K_m = Y_p(bh/L),$$ (9.75)

and the amplitude of its deflection is

$$|W| = (bh\sigma_1)/K_m.$$ (9.76)

The average power input is hence generally given by

$$\bar{P}_{in} = \frac{1}{2}\omega_r K_m |W|^2.$$ (9.77)

By substitution of Equations (9.75) and (9.76) in Equation (9.77), we get the aver-age power input in terms of the maximum applied stress:

$$\bar{P}_{in} = \frac{1}{2}\omega_r \frac{(\sigma_1)^2}{Y_p} bhL.$$ (9.78)

We then express Equation (9.78) in terms of beam power density to compare with the previous density found in Section 9.3:

$$\frac{\bar{P}_{\text{in}}}{Vol_{\text{b}}} = \frac{1}{2}\omega_{\text{r}}\frac{(\sigma_{1})^2}{Y_{\text{p}}}. \tag{9.79}$$

An input power density of 3.34 W/cm³ is obtained using our baseline parameters. This value is several orders of magnitude larger than the 154 µW/cm³ density obtained via Equation (9.69) for the mechanical power density based on the mass volume. This therefore suggests that, relatively speaking, a very large mass is needed to achieve such stress levels with the specified source and quality factor. In these conditions, the size of the device is therefore driven by its mass.

Let us now see how the electrical power density compares. Supposing that all the electrical energy is somehow harvested, we neglect the effect of the load in the complex power expression (Equation (9.38)). In this case, the electrical power is given by

$$\bar{P}_{\text{el}} = \frac{1}{2}\omega_{\text{r}}K_{\text{m}}|W|^2\kappa^2. \tag{9.80}$$

Due to the uniform stress field, the conversion factor is given by

$$\kappa^2 = \frac{k_{31}^2}{1 - k_{31}^2}. \tag{9.81}$$

For this block of PZT4, $\kappa^2 = 0.122$ and the suggested average electrical power density limit is 408 mW/cm³. Once more, despite a drop by an order of magnitude from the mechanical value, this electrical power density based on the beam stress limit is much larger than the 77 µW/cm³ value estimated from the mass volume and the vibration source input.

However, most resonators are not subject to a uniform stress field. Stress gradients during operation are commonly encountered and reduce the beam power density. This reduction is evaluated in the next subsection by considering the widely used cantilever beam geometry.

9.4.2 POWER DENSITY REDUCTION WITH STRESS DISTRIBUTION: THE CASE OF THE CANTILEVER

Let us assume that this same piezoelectric block is now subjected to a transverse force and behaves like a bending cantilever beam instead. The power expressions found in Equations (9.77) and (9.80) remain valid and two assumptions are made for simplicity:

- The piezoelectric conversion factor of the material, κ^2, specified in Equation (9.81) remains adequate to represent the conversion effectiveness of the transducer. This assumption is not accurate, but we will apply the necessary correction in Section 9.5.

- The stress distribution is that of a statically deformed cantilever beam with a constant rectangular section and a force applied at its free end. Hence, the static deflection shape is used and given by

$$\hat{w}(x) = \frac{1}{2}\left[-\left(\frac{x}{L}\right)^3 + 3\left(\frac{x}{L}\right)^2\right].$$ (9.82)

From the well known cantilever beam equations for deformation, stress, and stiffness (Equation (9.9)),

$$K_m = \frac{bh^3 Y_p}{4L^3},$$ (9.83)

$$\sigma_{max} = \frac{3}{2}\frac{hY_p}{L^2}|W|.$$ (9.84)

Replacing Equations (9.83) and (9.84) into Equation (9.77), the mechanical power input becomes

$$\bar{P}_{in} = \frac{1}{18}\omega_r \frac{(\sigma_{max})^2}{Y_p} bhL,$$ (9.85)

and the cantilever beam input power density can be written as

$$\frac{\bar{P}_{in}}{Vol_b} = \frac{1}{18}\omega_r \frac{(\sigma_{max})^2}{Y_p} = \frac{1}{9}\left(\frac{P_{in}}{Vol_b}\right)_{uniform}.$$ (9.86)

Comparing Equations (9.86) and (9.79), we conclude that an input power density reduction by a factor of 9 (to 371 mW/cm³) is due to the distribution of the strain along the length and across the thickness of the beam (see Figure 9.12). Given by

$$\frac{\bar{P}_{el}}{Vol_b} = \frac{1}{18}\omega_r \frac{(\sigma_{max})^2}{Y_p}\left(\frac{k_{31}^2}{1-k_{31}^2}\right),$$ (9.87)

the electrical power density of the cantilever is then equal to 45 mW/cm³. A triangular tapered beam, while stiffer, would maintain the stress over the entire length [25] and only have a reduction factor of 3 coming from the linear stress distribution through the thickness [13]. Hence, strictly from the mechanical strain distribution, the power density reduction in the beam is already significant.

Still, by comparison with the inertial limit established previously in Section 9.3, it is possible to estimate the mass to spring volume ratio for a similar power output from the ratio of Equations (9.87) and (9.71):

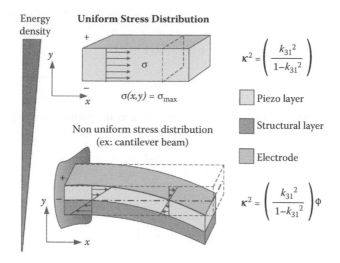

FIGURE 9.12 Effect of stress distribution on energy density.

$$\frac{Vol_M}{Vol_b} \propto \left(\frac{\omega_r^2}{\rho_M \mid A \mid^2 Q_m} \right) \left(\frac{\sigma_{max}^2}{Y_p} \right) \left(\frac{k_{31}^2}{1 - k_{31}^2} \right).$$ (9.88)

Despite the important input power density reduction observed for the cantilever geometry, the mass to beam volume ratio remains large at ≈584. Equation (9.88) indicates that for these two values to reach a similar level, one would have to improve Q_m and the density of the mass, ρ_M, significantly. Alternatively, a vibration source that is more than 20 times larger in magnitude would lead to similar volumes.

However, as stated before, the influence of the nonuniform stress distribution on the electromechanical conversion factor was not considered thus far, nor was the effect of the use of passive supporting layers. We analyze the impact of these elements in the next section, along with the influence of the piezoelectric material properties.

9.5 ELECTROMECHANICAL CONVERSION

Up to this point, we have assumed that the piezoelectric conversion factor, κ^2, was only dependent on the piezoelectric material (Equation (9.81)). However, this is only true for a bulk piezoelectric specimen that is uniformly stressed in a single direction. The conversion factor is reduced and can even reach small values approaching zero when a nonuniform stress is applied and passive layers are used. In this section, the effect of the geometry as well as the material properties is considered to provide a better assessment of the conversion effectiveness.

9.5.1 GEOMETRICAL CONSIDERATIONS

A convenient manner of expressing κ^2 consists of separating the piezoelectric properties from the geometrical and mechanical parameters [11,13]. Such a framework is useful for comparing and optimizing different configurations. First, we assume that, for a d_{31} mode device,

$$\theta = Y_p d_{31} \phi_1, \tag{9.89}$$

$$C_p = \varepsilon_{33}^S \phi_2, \tag{9.90}$$

where the variables ϕ_1 and ϕ_2 define the geometric contribution of each term. Based on Equations (9.6) and (9.7) from Section 9.2,

$$\phi_1 = \left(b\bar{h}_p \int_0^{L_1} \frac{d^2\hat{w}(x)}{dx^2} dx \right), \tag{9.91}$$

$$\phi_2 = \varepsilon_{33}^S \left(\frac{bL_1}{h_p} \right). \tag{9.92}$$

Equation 9.29 can then be rewritten as

$$\kappa^2 = \frac{d_{31}^2 Y_p^2}{\varepsilon_{33}^S} \frac{\phi_1^2}{\phi_2 K_m} = \frac{k_{31}^2}{1 - k_{31}^2} \phi, \tag{9.93}$$

with ϕ being the global geometric factor, expressed in the following equation:

$$\phi = \frac{Y_p \phi_1^2}{\phi_2 K_m}. \tag{9.94}$$

For simplicity, the static deflection shape (Equation (9.82)) is again used to evaluate the deformation dependent terms. Evaluating the integrals in Equations (9.91) and (9.9), we obtain

$$\int_0^{L_1} \frac{d^2\hat{w}(x)}{dx^2} dx = \frac{3}{2L} \left[-\left(\frac{L_1}{L} \right)^2 + 2\left(\frac{L_1}{L} \right) \right], \tag{9.95}$$

$$\int_0^L \left(\frac{d^2\hat{w}(x)}{dx^2} \right)^2 dx = \frac{3}{L^3}. \tag{9.96}$$

After some algebraic manipulations, ϕ for the unimorph beam can finally be expressed as [13]

$$\phi = \frac{9}{4} \frac{\tilde{Y}H(1-H)^2 \frac{L_1}{L} \left(2 - \frac{L_1}{L} \right)^2}{\left[(\tilde{Y}-1)H+1 \right]\left[(\tilde{Y}-1)^2 H^4 + (\tilde{Y}-1)H(4H^2 - 6H + 4) + 1 \right]}, \tag{9.97}$$

where $\tilde{Y} = Y_p/Y_s$ is the elasticity ratio between the active and structural layers, $H = h_p/h_{total}$ is the fraction of piezoelectric material that composes the beam and, L_1/L is the partial coverage ratio of the electrode.

Therefore, ϕ is a function of the elasticity ratio and the spatial distribution of the piezoelectric material on the structural layer. Equation (9.97) is plotted in Figure 9.13(a) for several elasticity ratios. The first thing noticeable is the invalidity of the previous assumption made in Section 9.4.2 regarding the conversion effectiveness of a bending beam made entirely from piezoelectric material. In fact, such a configuration would be completely ineffective, with $\phi = 0$. This occurs due to the charge cancellation between the bottom and top halves of the cantilever, which are stressed in opposite directions. Moreover, each elasticity ratio presents an optimal thickness fraction. For instance, with $\tilde{Y} = 1$ and $L_1/L = 1$, the form factor simplifies to $\phi = 9/4[H(1 - H)^2]$; the optimal thickness fraction is then $H = 0.33$, which gives $\phi_{opt} = 0.33$. This figure also demonstrates that coupling for the unimorph configuration is ideally increased by using a thick and compliant piezoelectric layer on a thin, rigid substrate. However, if we consider the thin film deposition techniques usually used in MEMS fabrication, the low fraction region is more relevant and, in this case, the trend rather favors a stiff and thin layer on a thick and compliant substrate.

Equation (9.98) is obtained by using the same simple scheme for a symmetric bimorph beam; the results are similarly illustrated in Figure 9.13(b):

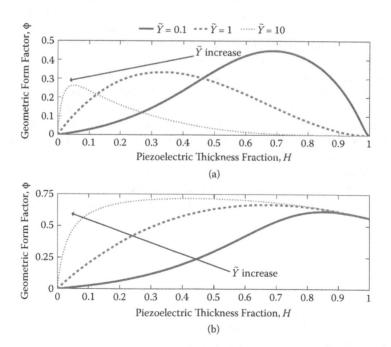

FIGURE 9.13 Form factor charts for various material stiffness ratios and thickness fraction for the (a) unimorph, and (b) bimorph configurations. $L_1/L = 1$.

$$\phi = \frac{9}{16} \frac{\tilde{Y}H\left[H^2 + 4(1-H)\right]\frac{L_1}{L}\left(2 - \frac{L_1}{L}\right)^2}{\left[(\tilde{Y}-1)H(H^2 - 3H + 3) + 1\right]}.$$ (9.98)

Contrary to the unimorph configuration, here $H = 2h_p/h_{total}$. We also note that the bimorph made exclusively from piezoelectric material does not suffer from the same charge cancellation problem, thanks to the segregation of the regions that are stressed in opposite directions. While the form factor remains far from unity, it is much better than for the unimorph with a value of ≈ 0.56. In general, the bimorph configuration is also better at conversion when stiffer and relatively thinner piezoelectric films are used. In both cases, using piezoelectric material near the neutral axis seems to decrease the effective coupling.

Several configurations, such as air-spaced [26] and tapered [27] cantilevers, have shown improved conversion factors as well. In fact, Figure 9.13(b) supports those claims in the case of air-spaced cantilevers since the curves tend to maximum asymptotic values as $\tilde{Y}$ increases. According to Yen et al., corrugated cantilevers can also be optimized to reach form factors that are similar to bimorph beams [28]. The advantages of this geometry are that it is easier to process than a bimorph while also eliminating the need for a passive support layer.

In addition, Equations (9.97) and (9.98) show that the geometric factor can also be optimized by restricting the length of the electrode on the beam. With our assumptions, the optimal coverage ratio is $L_1/L = 2/3$, which increases ϕ by about 19% (to 0.39 for a unimorph). In fact, the last third of the beam does not contribute as much to the conversion of energy due to the weakness of the bending moment in this portion. Hence, extending the electrode further contributes to an overall decrease of the conversion effectiveness. This charge redistribution phenomenon was recently validated experimentally by Stewart, Weaver, and Cain [29] and their results are consistent with the numbers we provide here. Therefore, the piezoelectric material should be located strategically on portions of the spring that are under large stresses to optimize the conversion effectiveness of the device.

Considering these results, the electromechanical conversion decreases from $\kappa^2 = 12.2\%$ (uniform stress) to $\kappa^2 = 4.8\%$ for an optimized unimorph cantilever (with $Y_p = Y_s$), which is about 2.5 times lower. The beam electrical power density, now expressed by

$$\frac{\bar{P}_{el}}{Vol_b} = \frac{1}{18}\omega_r \frac{(\sigma_{max})^2}{Y_p}\left(\frac{k_{31}^2}{1 - k_{31}^2}\right)\phi,$$ (9.99)

is equal to 18 mW/cm³, a value that is about 20 times less than what was first estimated for a uniformly stressed sample. Yet, it is still much larger than what we estimated from the inertial coupling limit. This again suggests that the density of the mass and the intensity of the acceleration source are the main drivers on the size of the device for typical ambient vibration sources.

TABLE 9.2

Global Geometric Factor for Various Configurations ($L_1/L = 1$)

Configuration	$\tilde{Y}$	H_{opt}	Φ_{opt}
Unimorph	1	1/3	1/3
Symmetric bimorph	1	2/3	2/3
Tapered unimorph	1	1/3	4/9
Tapered symmetric bimorph	1	2/3	8/9
Air spaced (symmetric)	∞	—	<0.75
Corrugated (optimized) [28]	1	—	2/3

Table 9.2 summarizes the assessments made in this section by collecting values of the global geometric factor estimated for various configurations. Note that the estimate for the air-spaced cantilever is based on the results for the bimorph by assuming $\tilde{Y} \rightarrow \infty$ and that the bending of the beam is not dominated by an S-shape—a possibility when the layers are very thin [26]. In this case, the electrode has to be segmented to avoid charge cancelation [30].

9.5.2 PIEZOELECTRIC MATERIAL CONSIDERATIONS

Even more important than geometry, the piezoelectric material properties have the dominant influence on the conversion effectiveness. Piezoelectric materials come in a variety of phases and crystallographic structures and are synthesized from a large range of processes. Each material and integration process presents its issues and challenges, affecting the cost as well as the performance of the device [31]. A complete review of these aspects is out of the scope of this chapter, but we provide here a brief overview of the potential candidates by considering how they are processed and their resulting properties. Tables 9.3 and 9.4 contain data collected for a wide variety of bulk, thick, and thin film piezoelectric materials.

It is also worth stressing some of the major differences between bulk piezoelectric materials and their thin film counterparts. First, the mechanical compliance of thin films is often dominated by the underlying thick substrate. Hence, they are subject to the influence of residual stresses due to mismatch in lattice constant and coefficient of thermal expansion differences, as well as coupling with other piezoelectric modes. Because these phenomena cannot be completely eliminated, it is difficult to measure the compliance of the film directly; rather, measurements lead to effective film properties. Most notably, the $e_{31,f}$ constant (in stress-charge notation) generally includes some coupling with the thickness mode. Equations (9.100) and (9.101) express the relationship of this constant with the d_{ij} piezoelectric constants in the strain-charge notation for two stress conditions [12,32,33]:

$$\text{Thin beam, main stress in one direction:} \quad e_{31,f} = \frac{d_{31}}{s_{11}^E}, \qquad (9.100)$$

TABLE 9.3

Reported Piezoelectric Properties for Bulk Materials

Material	Type	$(S_{11}^E)^{-1}$ (GPa)	$\varepsilon_{33}^T/\varepsilon_0$	d_{31} (pC/N)	k_{31} (%)	Q_m	Ref.
PZT-5A	Ceramic	61/71	1500/1928	−171/−190	34/37	75/80	15, 34, 35
PZT-5H	Ceramic	58/63	3200/3935	−250/−320	36/44	30/75	15, 34, 35
PZT-4	Ceramic	72/81	1135/1494	−120/−150	33/36	500/600	15, 34, 35
PZT-8	Ceramic	87/96	1000/1205	−93/−127	30/36	900/1050	15, 34–36
PMN-0.25PT	Ceramic	184	1167	−74	31	283	37
PMN-0.25PT	Single crystals	37.3	2560	−240/−569	33/73	131/362	37
PMN-0.29PT	Single crystals	—	5500	−1350	87	100	38
PMN-0.3PT	Single crystals	17.5/50	6610/7800	−742/−1395	49/90	44	39–41
PMN-0.32PT	Single crystals	15/34.8	650/5700	−160/−930	32/78	31/68	42
PMN-0.33PT	Single crystals	14.5	8200	−1330	59	—	43
PMN-0.345PT	Ceramic	75.4	4952	−255	75	—	44
PMN-0.42PT	Single crystals	106	660	−91	39	—	45
Soft PMN-PZT	Single crystals	7.87	4500/8000	−1400/−2252	90/95	100	38, 46
Hard PMN-PZT	Single crystals	—	3100/4000	−850/−1200	86/88	>500	38
PZN-0.045PT	Single crystals	9.35/28.6	2553/5600	−690/−1540	43/85	95/430	36, 47
PZN-0.045PT + 0.01Mn	Single crystals	13.2/43.5	1572/3491	−542/−830	42/80	375/441	47
PZN-0.045PT + 0.02Mn	Single crystals	27	1626/1873	−458/−502	69	336	47
PZN-0.07PT	Single crystals	14.8	3180	478	35	—	48
PZN-0.08PT	Single crystals	11.5	7700	−1455	60	40	41, 49
PZN-0.12PT	Single crystals	54	612.4	−148	50	—	44
BaTiO₃	Single crystals	124	168	−35	32/59	400	35, 39, 50
BaTiO₃	Ceramics	110/125	625/1700	−32/−78	15/21	300/1400	15, 34, 35
ZnO	—	127	11/12.64	−5.2/−5.43	18/19	N/A	15
LiNbO₃	—	173	30	−1	2.6	—	35

Thin plate, main stresses in two directions: $e_{31,f} = \dfrac{d_{31}}{s_{11}^E + s_{12}^E}$.   (9.101)

For this reason, an energy harvesting FOM, noted by $e_{31,f}^2/(\varepsilon_0\varepsilon_{33})$, is frequently used in the thin film processing literature instead of the usual coupling factor, k_{31}^2, that is used for bulk materials [32,33].

The data points in Tables 9.3 and 9.4 show that some materials provide interesting properties despite their much lower d_{31} constant. For example, barium titanate (BaTiO₃) offers comparable or better coupling factors than PZT ceramics because of its higher stiffness and lower dielectric constant. Thin film aluminum nitride (AlN) also compares favorably. However, we also want to stress that considering materials

TABLE 9.4
Reported Piezoelectric Properties for Some Thin Film Materials

Material	Type	Substrate	$\varepsilon_{33}^S/\varepsilon_0$	$e_{31,f}$ (C/m²)	$e_{31,f}^2/\varepsilon_0\varepsilon_{33}$ (GPa)	Q_m	Ref.
AlN	MOCVD Epitaxial	Sapphire	9.5	−1.37	22.3	2490	51
AlN	Sputtering	Si	10.2	−1.3	18.7	–	51
AlN	Sputtering	—	10.5	−1.05	11.9	—	32
AlN	Sputtering	Si	—	—	—	120/500	52, 53
ZnO	Sputtering, single crystal	Si	10.9	−1.0	10.3	1770	32, 51, 54
PZT (53/47)	Single crystal, sputtering	MgO	200	−6.2	21.7	—	55
PZT (53/47)	Polycrystalline, sputtering	Si	200	−7.7	33.48	—	55
PZT	Sol-gel	SiO₂/Si	1100/1300	−4/−6	6/18	54/237	51, 56
PZT (48/52)	—	MgO	—	−3.98	—	114	57
PMnN-0.94PZT (48/52)	Single crystal, sputtering	MgO	100	−12.0	163	185	57
PMnN-0.94PZT (50/50)	Polycrystaline, sputtering	SiO₂/Si	834	−14.9	30	—	57
PMN-0.33PT	Single crystal, sputtering	MgO	500	−5	5.65	20	58
PMN-0.33PT	Single crystal, sputtering	SrTiO₃/Si	1600	−27	50	—	59

on the unique basis of their piezoelectric properties can be misleading. Indeed, materials with a very high coupling factor are commonly assumed to be highly desirable for high-performance energy harvesters (e.g., single crystal relaxers). Even if Equation (9.99) does seem to support this claim, we must also consider the conclusions relative to the inertial coupling limit established in Section 9.3. This limit states that an increase of the coupling beyond a critical point does not provide benefits on the net power output or power density of the resonator due to excessive damping, which reduces the captured energy. Because this limit is also closely tied to mechanical damping, it is important to consider the material mechanical quality factor, Q_m, as well. Later in this chapter, we discuss the importance of this property in the selection process by also considering the impact of the electric load. For that matter, the specific topic of the energy extraction circuit is addressed in the following section.

9.6 ELECTRICAL ENERGY EXTRACTION

To be harvested properly, the energy converted into electricity by the transducer must be extracted by the connected load. Therefore, the performance of an electrical energy extraction interface could be considered in terms of *effectiveness* and *efficiency*. The first considers the fraction of converted energy extracted and the second the fraction of successfully extracted energy, accounting for losses like Joule effect or current leakage, for instance. We will briefly touch on the effectiveness aspect of the extraction process in this section.

9.6.1 Resistive Load

In Section 9.2 we presented how the piezoelectric element can be modeled as a current source connected in parallel to a capacitance, C_p. Based on this approach, we then derived the actual output power, $\bar{P}_{el}$, from the active power delivered to the resistive load. A closer look at the complex power expression (Equation 9.38) is, however, enough to show that the impedance-matched resistance enables half of the electric energy converted to be harvested continuously at the steady state.

Indeed, the first part is the electrical power expressed in terms of the electrical stiffness ($K_m \, \kappa^2 = K_{el}$ for an open circuit), while the second term modulates this power as a function of operating frequency and load. The proportions of real and reactive powers are also included in this term. For a short circuit ($\Omega\alpha = 0$), the modulus of the second term is zero and there is no power generated. For an open circuit ($\Omega\alpha \rightarrow \infty$), the modulus of this second term is now one, but all the generated power is reactive. In contrast, a matched impedance load leads to an even distribution of active and reactive power, but the modulus of the second term is slightly lowered to $\sqrt{1/2}$. Therefore, the maximum extraction effectiveness of a crude matched resistance is indeed 50% for steady-state resonance operation.

However, the raw output of this simple interface is generally AC current. To store or use this energy, it must first be rectified, filtered, and regulated appropriately. Doing so requires the use of diodes and storage capacitances, introducing losses and threshold voltages. The losses affect the global efficiency, but these threshold voltages also block the energy from flowing out of the transducer, which impacts the effectiveness of the interface. It is also possible to use nonlinear extraction schemes to improve the energy transfer and the harvesting efficiency.

9.6.2 Nonlinear Extraction Interfaces

A practical energy harvester requires the implementation of an energy buffer in order to assure a reliable operation under various circumstances. Capacitances, supercapacitances, or batteries are often necessary because of the fluctuation of the available ambient energy. The energy generation and consumption are also not always synchronized [1]. Although this makes the system complex, it allows the use of energy extraction circuits between the harvester and the storage to improve extraction effectiveness. In some cases, the storage stage may not be directly connected to the piezoelectric transducer and this also affects the extraction process [60]. Another significant implication is that nonlinearities can be used to synchronize the extraction of energy with the maximum displacement of the transducer, which somewhat decouples the energy injection and extraction process. For example, the synchronous electric charge extraction (SECE) and double synchronized switch harvesting (DSSH) architectures allow a decoupling of the extraction and storage stages [61].

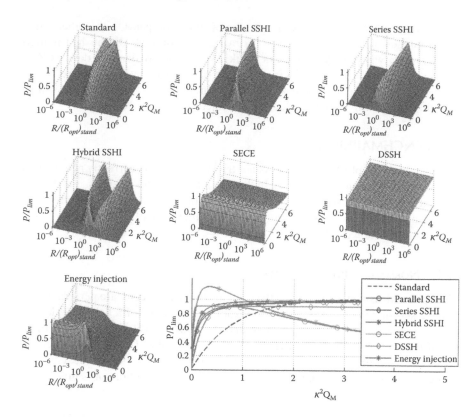

FIGURE 9.14 Normalized harvested power as a function of load and normalized maximum power for several extraction interfaces under a constant force magnitude. (From Lallart, M., and Guyomar, D. 2011. *IOP Conference Series: Materials Science and Engineering* 18 (9): 092006.)

Because these nonlinear harvesting interfaces improve the effectiveness of the extraction process, they induce more damping on the structure for a given coupling level compared to a simple shunt resistance. However, we have shown in Section 9.3.1 that a backward coupling effect can limit the net output power. We have also shown in Section 9.2.4 that the corresponding resonator FOM for which this backward effect occurs with a resistive load is $\kappa^2 Q_m = 2$, but this value changes for different circuits.

The plots on Figure 9.14, adapted from Lallart and Guyomar [61], show that every interface tends toward the same limit, P_{lim}, with the increase of $\kappa^2 Q_m$. Nonetheless, the nonlinear architectures reach this power limit for lower values of the resonator FOM, indeed suggesting an improvement of the extraction effectiveness. The only exception is the injection technique, which actually surpasses the inertial limit. However, this comparison is not entirely fair, because electrical energy is added in the system [60].

Hence, the benefits of these circuits are significant for highly damped or lightly coupled devices that have a low resonator FOM ($\kappa^2 Q_m \ll 2$). In some cases, gains of more than 400% were observed in terms of power output. On the other hand, the advantages of these very effective extraction schemes are diminished if used in conjunction with highly coupled or weakly mechanically damped devices [62], because they cannot overcome the inertial limit stated by Equation (9.71). Yet, these interfaces can improve both the harvesting efficiency and the bandwidth [63], which, as we discuss later in Section 9.8.1, can be of interest in some scenarios.

9.7 BENCHMARKING

The benchmarking of PVEH devices is not trivial and there is currently no standard framework or procedure for doing so. In this section, we propose, review, and discuss different approaches based on the different limits that were exposed. The proposed FOMs, summarized in Table 9.5, are briefly explained before being used to assess the performance of recently reported devices.

9.7.1 Normalized Power Density

While power density has so far been the usual metric used to compare performances, we have shown here that it only partially captures the value of a harvester design. Moreover, Equations (9.64)–(9.71) show that the power density is not only a function of the properties of the PVEH device (like the quality factor Q_m, the density of the mass ρ_M, or the piezoelectric coupling), but it is also tied to the vibration source (described by its amplitude $|A|$ and frequency ω). Therefore, it is misleading

---

TABLE 9.5
Summary of the Figures of Merit

FOM	Expression	Purpose
Normalized power density	$P_\rho = P_{el}\,\omega/(A^2 Vol)$	Normalizes the power output by the specific power of the single frequency source vibration and the device volume; used to compare devices designed for different sinusoid sources; expressed like a mass density (e.g., in grams per cubic centimeter)
Piezoelectric resonator	$\kappa^2 Q_m$	Compares mechanical losses to the electromechanical coupling of the structure; this dimensionless FOM is used to compare the levels of mechanical damping and electrical damping
Harvesting efficiency	$\eta = \dfrac{P_{el}}{P_{in}} = \dfrac{\zeta_{el}}{(\zeta_m + \zeta_{el})}$	Gives the fraction of energy pumped in the device that is used to produce power
Harvesting effectiveness	$\xi = \dfrac{P_{el}}{P_{lim}} = 4\dfrac{\zeta_{el}}{(\zeta_m + \zeta_{el})^2}$	Indicates if the device is close to the maximum power limit imposed by inertial coupling for a single frequency sinusoid source
Beam energy density	$E^* = P_{el}/Vol_b f$	Evaluates if the piezoelectric spring element is highly stressed

to use this value as a metric to compare designs operating under different conditions because the intensity of the vibration applied is included.

Instead of strictly using the power density, Beeby et al. [64] proposed its normalization by the square of the acceleration amplitude, $|A|^2$. However, this normalization does not capture the effects of the frequency of the source. To find the solution to this problem, one simply needs to take into account the input power (Equation 9.66) by instead considering normalizing by $|A|^2/\omega$. Expressed in watts per kilogram (or square meters per cubic second), it can be viewed as a specific power that characterizes the quality of the source in some manner. Therefore, an adequate normalized power density FOM, P_ρ, can be expressed as [18,53]

$$P_\rho = P_{el}\omega / (|A|^2 \, Vol) \tag{9.102}$$

$$= \frac{M_{eq}}{4Vol} \frac{\zeta_{el}}{(\zeta_m + \zeta_{el})^2} \tag{9.103}$$

$$= \frac{M_{eq}}{2Vol} Q_{tot}\eta. \tag{9.104}$$

Here, $Vol = Vol_b + Vol_M$ is the total effective volume of the transducer with its tip mass. Therefore, the units of P_ρ are similar to a density and it scales with ρQ_m to enable a fair comparison of the power output from different vibration conditions. The use of a similar FOM was also proposed by Mitcheson et al. [18], who suggested normalizing the output power by $Y_0^2\omega^3 M_{eq}$, where Y_0 is the source vibration displacement. However, we propose using the volume instead of the mass to reward designs that integrate high-density materials. At the same time, normalizing by the volume instead of the mass provides an indication of how much of the whole transducer inertia is useful in capturing energy.

9.7.2 Piezoelectric Resonator FOM: Harvesting Efficiency and Effectiveness

We have discussed how the piezoelectric resonator FOM, $\kappa^2 Q_m$, can also be a very useful metric to compare devices due to its ties to the resonator harvesting efficiency, η (Equations (9.72) and (9.73)) and its harvesting effectiveness, ξ (Equation (9.74)). These two other FOMs were discussed in Section 9.3 of this chapter. The first indicates the fraction of the energy pumped in the device used to produce power, while the second states whether the device is close to the limit imposed by inertial coupling. Because both are affected by the electrical damping, they also depend on the load.

9.7.3 Beam Energy Density

Even though a design might be capable of providing a very large power density, it might be at the expense of its reliability. Therefore, the induced stresses should be considered in the benchmarking procedure to evaluate if the piezoelectric spring element is used or sized appropriately. Based on the discussions presented in Sections 9.4 and 9.5, a fourth FOM should be introduced to evaluate the energy density of the beam. We propose to define this FOM as $E^* = P_{el}/Vol_b f$ and expect it to scale with coupling, κ^2, and stress, σ. Therefore, we also estimate the stress to check this assumption.

We cannot directly use Equation (9.99) to evaluate the stress because it was developed for a homogeneous cantilever beam; actual devices use composite structures. Equation (9.80) cannot be used due to the lack of information in the experimental literature concerning either the beams' stiffness or their amplitude of displacement. Rather, we deduce the maximum strain energy from the electrical power output, the coupling factor, and the load. Indeed, from the definition of the quality factor (Equation (9.47)), it is possible to say that

$$E_{strain} = \frac{P_{el}Q_{el}}{\omega_r}. \tag{9.105}$$

We then use the provided geometric parameters to estimate the bending stiffness of the beams, YI, and evaluate an RMS (root mean square) curvature based on the resulting strain energy,

$$E_{strain} = \frac{1}{2} YI \int_0^L \left(\frac{\partial^2 w}{\partial x^2} \right)^2 dx \tag{9.106}$$

$$\left. \frac{\partial^2 w}{\partial x^2} \right|_{RMS} = \left[\frac{1}{L} \int_0^L \left(\frac{\partial^2 w}{\partial x^2} \right)^2 dx \right]^{1/2} = \left(\frac{2E_{strain}}{YIL} \right)^{1/2}. \tag{9.107}$$

An estimate of the mean stress at the top of the piezoelectric layer is finally obtained as

$$\sigma = Y_p h_c \left. \frac{\partial^2 w}{\partial x^2} \right|_{RMS}. \tag{9.108}$$

9.7.4   DEVICE ASSESSMENTS

The proposed FOMs were used to compare different PVEH resonators reported in the literature. The results of this investigation are summarized in Table 9.6. All these devices produce power on the order of 1 to 50 µW.

As previously noted, P_p does scale with ρQ_m, motivating the choice of high-density materials and devices with high Q_m. For example, Renaud and colleagues' device [67], with its very high Q_m, has a P_p larger than all the others. However, a low value of Q_m can be somewhat compensated by using materials with higher density, as can be seen from the device reported in Morimoto et al. [66]. Using stainless steel as the structural layer instead of silicon, this device can capture more inertial energy due to an increased inertial force. In terms of maximum efficiency, η_{max}, the devices made

TABLE 9.6

Assessment of Several PVEH Devices Reported in the Literature

	Ref.	65	66[a]	67	46a	IMEC 1A[b]	IMEC 4B[b]	IMEC 7A[b]
Design	Piezo material	d_{33} PZT	d_{31} PZT	d_{31} AlN	d_{31} PMN-PT	d_{31} AlN	d_{31} AlN	d_{31} AlN
	Substrate	Silicon	Stainless steel	Silicon	Aluminum	Silicon	Silicon	Silicon
	Process	Sol-gel	Epitaxial	—	Bonded bulk	—	—	—
	Piezo thickness (μm)	2	2.8	1.2	500	2	2	2
	Beam thickness (μm)	9	56	51.2	1330	52	52	52
	Thickness Fraction	0.22	0.05	0.023	0.39	0.038	0.038	0.038
	Mass[c] (mg)	0.76	38	14	490	15	15	40
	Mass material	Silicon	—	Silicon	—	Silicon	Silicon	Silicon
	Vol_b (mm³)	0.0029	5.1	0.061	106	0.22	0.27	0.37
	Vol_M (mm³)	0.32	0	6	0	6	6	16.9
	f (Hz)	877	126	1082	≈1835	1178	1002	620
	A (m/s²)	19.6	5	3.14	9.81	6.28	6.28	6.28
	A^2/ω (m²/s³)	0.070	0.032	0.0015	0.0083	0.0053	0.0063	0.010
Performance	Q_m	220[c]	85	1200	22	1900	1450	865
	k_e^2	0.05	0.013	0.0025	0.18	0.0039	0.004	0.003
	$\kappa^2 Q_m$	11.6[a]	1.12	3	4.83	7.44	5.82	2.60
	σ^c (MPa)	84	7	170	92	163	158	459
	P_{el} (μW)	1.4	5.3	3	7.3	10	10	32
FOM	P_{el}/Vol (μW/cm³)	4255	1039	249	69	1608	1595	1853
	P_ρ (g/cm³)	62	33	333	8.3	297	250	186
	η_{max} (%)	85	36	60	71	79	74	57
	ξ	0.98	0.67	1	1	0.53	0.58	0.74
	E^* (nJ/mm³)	550	8.3	46	0.04	38	37	87

[a] Calculation corrected for lack of tip mass.
[b] Operating at maximum efficiency.
[c] Estimated values.

of AlN fabricated at IMEC all do very well compared to the other devices reported, despite their relatively low coupling, showing the benefits of their high-quality factor. Only the device reported in Muralt et al. [65] is better, because of the combination of its good Q_m and high coupling factor. The latter is achieved with an interdigitated electrode design that benefits from the better d_{33} mode properties.

As expected, highly stressed IMEC 7A and highly coupled [65] beams present higher values of E^*. Although increasing coupling will not increase the power density based on the mass volume (P_ρ), it will reduce the beam or piezoelectric volume for a desired electrical power. In addition, Table 9.6 clearly shows that devices with $\kappa^2 Q_m \geq 2$ have been frequently fabricated at the MEMS scale and even some at the mesoscale [46], which contradicts the claims of Guyomar and Lallart that realistic devices usually have $\kappa^2 Q_m$ values not much higher than 0.2 [60]. Therefore, multiple devices that reach the inertial coupling limit have been realized in recent years.

9.8 OTHER CONSIDERATIONS

This chapter focused primarily on linear resonant PVEHs operating under a single frequency sinusoid excitation. In this last section, however, we broaden the scope with a discussion on wide band vibration energy harvesting. More specifically, we introduce some nuances concerning wideband vibration sources and nonlinear approaches for vibration energy harvesting. We also discuss potential design trade-offs and their impacts on the choice of the piezoelectric material.

9.8.1 FREQUENCY DEPENDENCE OF THE INERTIAL COUPLING LIMIT

An important limit of resonant devices is that they must operate at a specific frequency. Moreover, we showed that a high Q factor is required to improve the power density, but this implies a small operation bandwidth. If the mechanical energy source is broadband or nonstationary in nature, which is the case for most ambient vibrations, the harvester may collect only a small portion of the available energy or, worse, have a very low power output and a poor power density due to its inability to match its frequency with the source. Reducing the quality factor improves the bandwidth, but it might be at the expense of the power output. Addressing this problem is an active area of current research.

For a single frequency source, the optimal power is harvested by matching the electrical damping with the mechanical damping. However, the optimization is not as straightforward for broader band sources [21]. Figures 9.4 and 9.5 show the power output frequency response changes with the different damping type. If only the mechanical damping increases, the power at the resonant peak is reduced, but remains largely the same outside the peak. Therefore, the bandwidth increase only comes from a diminution of the main peak and the average output power is lower. However, by instead increasing the electrical damping, it is possible to improve the frequency response outside the peak. Indeed, while the peak may be lower, the response for the nearby frequencies is increased.

This general broadening of the response can be sufficient to compensate for the main peak diminution, but this depends on the shape of the power spectral density (PSD) of the harvested vibration. Hence, for wider band sources, matching the damping components is not necessarily the optimal scenario and improving the efficiency is preferable [21,68]. Nonetheless, increasing the electrical damping to improve the bandwidth has its limits due to the observed reduction of the resonant peak. Other approaches are being investigated to increase the bandwidth with less peak attenuation.

A simple solution is to use multimodal structures or even arrays of piezoelectric structures sensitive to different frequencies [69]. This approach broadens the operational bandwidth but at the expense of the volumetric power density. Several papers have proposed methods to tune the resonant frequency mechanically [70] or electrically [71] to provide broadband operation while limiting the total size of the device. Nevertheless, active (continuous) or semiactive (discontinuous) tuning requires energy consumption and it is not clear if the net energy balance can be improved significantly by these schemes. Hence, passive approaches are also being investigated to introduce instabilities and other nonlinearities to improve the mechanical power transfer to the transducer. For example, devices incorporating nonlinear compliant structures [72–77], magnetic coupling [78–80], or stoppers [81–85] have been the topic of multiple papers in recent years.

9.8.2 POSSIBLE COMPROMISE CONSIDERING MATERIALS AND EXTRACTION CIRCUITS

To provide a broad perspective on the selection of piezoelectric materials, we mapped the quality factor Q_m with estimates of κ^2 on Figure 9.15. This figure considers the 31 mode properties for the materials that are reported in Tables 9.3 and 9.4, without the effect of the geometry of the composite structure (i.e., a geometric form factor $\phi = 1$ is assumed). For comparison, the devices benchmarked in Section 9.7.4 are also added to this figure. The dashed line is the condition $\kappa^2 Q_m = 2$, which corresponds to the critical coupling value with a resistive load. We note that most of the materials and reported devices are located in the optimal design region, which suggests that coupling should not be the only criterion for PVEH material selection.

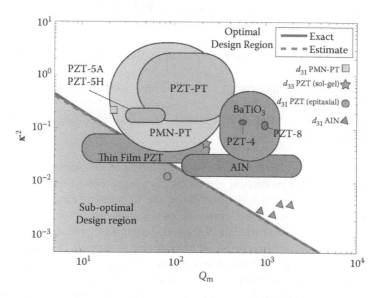

FIGURE 9.15 Critical coupling factor with respect to Q_m and mapping of materials for this criterion.

By using nonlinear extraction interfaces, it is also possible to improve the performance of designs that have a lower resonator FOM and, therefore, allow multiple design trade-offs. For instance, with a high-coupling material and a highly effective extracting interface, the size of the piezoelectric active element can be reduced significantly [37,86]. Alternatively, the use of more effective extraction schemes allows a wider selection of piezoelectric materials (e.g., low-coupling materials) to produce the same power output. Hence, many application-specific factors can be considered for the selection of materials, including:

- Processing of the materials: time, costs, or design requirements (e.g., integrated circuits, wafer level packaging process compatibility)
- Toxicity (e.g., Restriction of Hazardous Substances compliance, biocompatibility for bioMEMS)
- Curie temperature (no degradation during fabrication or usage)
- Dielectric losses and leakage behavior
- Device operating principle (resonant or nonresonant) and characteristics of vibrations in the considered application (magnitude and type, e.g., random, shocks, harmonics, colored noise, etc.)
- Fatigue strength (concern for harsh environments)

9.9 CONCLUSION

This chapter presented a discussion on the various aspects that can limit the power density of resonant piezoelectric vibration energy harvesters. To support the analyses, we introduced a streamlined, lumped parameter modeling approach. To address the need for relevant metrics of comparison for PVEH system design, we also investigated several ways to assess the performance of devices:

1. Compare the amount of energy captured by the device with the quality of the source.
2. Evaluate the amount of energy converted into electricity by the transducer and its energy density.
3. Evaluate the amount of electric energy extracted from the device and its global efficiency.

All these points were reviewed for linear resonators by considering the impact of the geometry, material properties, and electrical energy extraction interface used. For typical values, a power density limit below the 100 µW/cm³ range was estimated due to the inertial coupling limit. However, to reach this power density, the piezoelectric spring must be well designed. A good design must provide sufficient coupling not to limit the achievable power output and to support the induced stress without failure. However, we have shown that for low amplitude vibrations, capturing a significant amount of energy is the most important bottleneck, because of the large mass required compared to the size of the beam. Therefore, the power density will be first limited by the mass density of the device and then by stress.

Nonetheless, if very large accelerations or forces are available in the ambient environment (e.g., in harsh environments), the strength of the materials may in this case become a concerning factor. For designs that provide low coupling, nonlinear extraction interfaces can be alternatively used to improve the performance. The inertial coupling limit, however, limits the gain in performance for devices that present high Q_m.

Based on this investigation, we proposed alternative FOMs such as the normalized power density, the harvester's efficiency, and the harvester's effectiveness. Interestingly, all of these depend on the piezoelectric resonator FOM, $\kappa^2 Q_m$. The beam energy density was also proposed as a new FOM to evaluate the beam stress and thus an indirect method to assess its design and guide the selection of both materials and geometry to achieve a specified power level in a compact beam volume. An assessment of several resonant devices recently published in the literature was then conducted, revealing that many have already reached the inertial limit expected for resonators. Moreover, the devices lagging behind could reach this limit with some geometric optimization or with more effective extracting interfaces. As expected, devices with a large quality factor presented the highest normalized power density values. However, the travel of the mass was not considered in our estimates. Taking this into account would reduce further the power density [18]. While these metrics provide a framework for comparing inertial piezoelectric energy harvesters, further reflections are required to define FOMs that account for broadband operation, costs, and fabrication considerations. Equivalent FOMs for nonresonant devices would also be welcomed as a means to compare different operating principles.

This global analysis suggests that the technology for piezoelectric generator energy conversion is quite developed. Research on processing of very high coupling materials is well underway and analytical tools for the design and optimization of piezoelectric transducers have also been developed. Concerning the energy extraction interfaces, an array of circuit topologies has been proposed, but their implementation and performance at the microscale pose unresolved challenges. However, the power density of PVEH devices is inherently limited by the power input and, consequently, the methods that are currently used to capture the ambient energy. It is assumed that a very large quantity of vibrational energy is available in the environment, but it remains challenging to harness energy from such unpredictable, low-quality sources. Linear, resonant harvesters appear to have reached their theoretical limits and significant research efforts are underway to bring the same level of performance to a broader range of operating frequencies and scenarios. Frequency tuning methods have been proposed and novel nonlinear techniques to transfer energy from one frequency band to another should also be researched. This challenge has been tackled for a couple of years, but no clear winning solution has emerged yet. In fact, it is quite probable that no single solution can address every type of solicitation.

ACKNOWLEDGMENTS

This research was supported by General Motors of Canada and by the Natural Sciences and Engineering Research Council (NSERC) of Canada through the

Canada Research Chairs, Post Graduate Scholarship, and CREATE ISS programs. We would also like to acknowledge the cooperation of Ruud Vullers and Rene Elfrink at IMEC, as we are grateful for the PVEH devices data that they provided for our assessments.

REFERENCES

1. Vullers, R. J. M., van Schaijk, R., Doms, I., Van Hoof, C., and Mertens, R. 2009. Micropower energy harvesting. *Solid-State Electronics* 53 (7): 684–693.
2. Beeby, S. P., Tudor, M. J., and White, N. M. 2006. Energy harvesting vibration sources for microsystems applications. *Measurement Science and Technology* 17:R175.
3. Erturk, A., and Inman, D. J. 2008. Issues in mathematical modeling of piezoelectric energy harvesters. *Smart Material Structure* 17 (6): 065016.
4. Erturk, A., and Inman, D. J. 2008. A distributed parameter electromechanical model for cantilevered piezoelectric energy harvesters. *Journal of Vibration and Acoustics—Transactions of the ASME* 130 (4): 041002.
5. duToit, N. E., Brian, L. W., and Kim, S. G. 2005. Design considerations for MEMS-scale piezoelectric mechanical vibration energy harvesters. *Integrated Ferroelectrics* 71 (1): 121–160.
6. Williams, C. B., and Yates, R. B. 1996. Analysis of a micro-electric generator for microsystems. *Sensors and Actuators A: Physical* 52 (1–3): 8–11.
7. Roundy, S., Wright, P. K., and Rabaey, J-M. 2004. *Energy scavenging for wireless sensor networks: With special focus on vibrations.* Norwell, MA: Kluwer Academic Publishers.
8. Roundy, S. 2005. On the effectiveness of vibration-based energy harvesting. *Journal of Intelligent Material Systems and Structures* 16 (10): 809–823.
9. Lallart, M., and Guyomar, D. 2010. Piezoelectric conversion and energy harvesting enhancement by initial energy injection. *Applied Physics Letters* 97:014104.
10. Tabesh, A., and Fréchette, L. G. 2009. On the concepts of electrical damping and stiffness in design of a piezoelectric bending beam energy harvester. *Proceedings of Power MEMS 2009,* Washington, DC, December 1–4, 368–371.
11. Dompierre, A., Vengallatore, S., and Fréchette, L. G. 2010. Compact model formulation and design guidelines for piezoelectric vibration energy harvesting with geometric and material considerations. In *Proceedings of the 10th Workshop on Micro and Nanotechnology for Power Generation and Energy Conversion Applications—PowerMEMS 2010* December 1–3, Leuven, Belgium, technical digest poster sessions, 99–102, Micropower Generation Group, imec/Holst Center.
12. Erturk, A. 2009. Electromechanical modeling of piezoelectric energy harvesters. PhD thesis, Virginia Polytechnic Institute and State University, Department of Engineering Science and Mechanics, Blacksburg, VA.
13. Dompierre, A. Modélisation et conception de microrésonateurs piézoélectriques pour la récupération d'énergie vibratoire. Master's thesis, Université de Sherbrooke, Mechanical Engineering Department, Sherbrooke, Quebec, August 2011.
14. Standards Committee of the IEEE Ultrasonics, Ferroelectrics, and Frequency Control Society. ANSI/IEEE Std 176-1987: Standard on piezoelectricity, January 1988.
15. Jaffe, H., and Berlincourt, D. A. 1965. Piezoelectric transducer materials. *Proceedings of the IEEE* 53 (10): 1372–1386.
16. Mandal, N. K., and Biswas, S. 2005. Vibration power flow: A critical review. *Shock and Vibration Digest* 37 (1): 3.

17. Antonio, J. 1984. Power flow in structures during steady-state forced vibration. PhD thesis, Imperial College of Science and Technology, London, UK.
18. Mitcheson, P. D., Yeatman, E. M., Rao, G. K., Holmes, A. S., and Green, T. C. 2008. Energy harvesting from human and machine motion for wireless electronic devices. *Proceedings of the IEEE* 96 (9): 1457–1486.
19. Richards, C. D., Anderson, M. J., Bahr, D. F., and Richards, R. F. 2004. Efficiency of energy conversion for devices containing a piezoelectric component. *Journal of Micromechanics and Microengineering* 14:717.
20. Shu, Y. C., and Lien, I. C. 2006. Efficiency of energy conversion for a piezoelectric power harvesting system. *Journal of Micromechanics and Microengineering* 16:2429.
21. Renaud, M., Elfrink, R., Jambunathan, M., de Nooijer, C., Wang, Z., Rovers, M., Vullers, R., and van Schaijk, R. 2012. Optimum power and efficiency of piezoelectric vibration energy harvesters with sinusoidal and random vibrations. *Journal of Micromechanics and Microengineering* 22 (10): 105030.
22. Izyumskaya, N., Alivov, Y. I., Cho, S. J., Morkoc, H., Lee, H., and Kang, Y. S. 2007. Processing, structure, properties, and applications of PZT thin films. *Critical Reviews in Solid State and Materials Sciences* 32 (3–4): 111–202.
23. PI. http://www.physikinstrumente.com/tutorial/4_22.html
24. Cain, M. G., Stewart, M., and Gee, M. G. 1999. Degradation of piezoelectric materials. National Physical Laboratory Management Ltd., Teddington, Middlesex, UK, NPL Rep. SMMT (A), 148.
25. Goldschmidtboeing, F., and Woias, P. 2008. Characterization of different beam shapes for piezoelectric energy harvesting. *Journal of Micromechanics and Microengineering* 18 (10): 104013.
26. Wang, Z., and Xu, Y. 2007. Vibration energy harvesting device based on air-spaced piezoelectric cantilevers. *Applied Physics Letters* 90 (26): 263512–263512.
27. Halvorsen, E., and Dong, T. 2008. Analysis of tapered beam piezoelectric energy harvesters. In *Proceedings of PowerMEMS 2008 + microEMS 2008*, Sendai, Japan, November 9–12, 241–244.
28. Yen, T. T., Hirasawa, T., Wright, P. K., Pisano, A. P., and Lin, L. 2011. Corrugated aluminum nitride energy harvesters for high energy conversion effectiveness. *Journal of Micromechanics and Microengineering* 21:085037.
29. Stewart, M., Weaver, P. M., and Cain, M. 2012. Charge redistribution in piezoelectric energy harvesters. *Applied Physics Letters* 100 (7): 073901–073901.
30. Erturk, A., Tarazaga, P. A., Farmer, J. R., and Inman, D. J. 2009. Effect of strain nodes and electrode configuration on piezoelectric energy harvesting from cantilevered beams. *Journal of Vibration and Acoustics* 131 (1): 011010.
31. Vullers, R. J. M., van Schaijk, R., Goedbloed, M., Elfrink, R., Wang, Z., and Van Hoof, C. 2011. Process challenges of MEMS harvesters and their effect on harvester performance. In *2011 IEEE International Electron Devices Meeting (IEDM)*, 10.2.1–10.2.4, December.
32. Trolier-McKinstry, S., and Muralt, P. 2004. Thin film piezoelectrics for MEMS. *Journal of Electroceramics* 12:7–17.
33. Trolier-McKinstry, S., Griggio, F., Yaeger, C., Jousse, P., Zhao, D., Bharadwaja, S. S. N., Jackson, T. N., Jesse, S., Kalinin, S. V., and Wasa, K. 2011. Designing piezoelectric films for micro electromechanical systems. *IEEE Transactions on Ultrasonics, Ferroelectrics and Frequency Control* 58 (9): 1782–1792.
34. Matweb. http://www.matweb.com/
35. eFunda. http://www.efunda.com/materials/piezo/material_data/matdata_index.cfm

36. Lebrun, L., Sebald, G., Guiffard, B., Richard, C., Guyomar, D., and Pleska, E. 2004. Investigations on ferroelectric PMN-PT and PZN-PT single crystals ability for power or resonant actuators. *Ultrasonics* 42 (1–9): 501–505.

37. Badel, A., Benayad, A., Lefeuvre, E., Lebrun, L., Richard, C., and Guyomar, D. 2006. Single crystals and nonlinear process for outstanding vibration-powered electrical generators. *IEEE Transactions on Ultrasonics, Ferroelectrics, and Frequency Control* 53 (4): 673–684.

38. Ceracomp. http://www.ceracomp.com/

39. Peng, J., Luo, H., He, T., Xu, H., and Lin, D. 2005. Elastic, dielectric, and piezoelectric characterization of $0.70Pb(Mg_{1/3}Nb_{2/3})O_3$-$0.30PbTiO_3$ single crystals. *Materials Letters* 59 (6): 640–643.

40. Zhang, R., Jiang, W., Jiang, B., and Cao, W. 2002. Elastic, dielectric and piezoelectric coefficients of domain engineered $0.70Pb(Mg_{1/3}Nb_{2/3})$ O_3-0.30 $PbTiO_3$ single crystal. In *AIP Conference Proceedings*, 188–197. IOP Institute of Physics Publishing LTD.

41. Park, S. E., and Shrout, T. R. 1997. Characteristics of relaxor-based piezoelectric single crystals for ultrasonic transducers. *IEEE Transactions on Ultrasonics, Ferroelectrics and Frequency Control* 44 (5): 1140–1147.

42. APC International Ltd. http://www.americanpiezo.com/product-service/pmn-pt.html

43. Zhang, R., Jiang, B., and Cao, W. 2001. Elastic, piezoelectric, and dielectric properties of multidomain $0.67Pb(Mg_{1/3}Nb_{2/3})O_3$-$0.33PbTiO_3$ single crystals. *Journal of Applied Physics* 90:3471.

44. Delaunay, T., Le Clézio, E., Guennou, M., Dammak, H., Thi, M. P., and Feuillard, G. 2008. Full tensorial characterization of PZN-0.12PT single crystal by resonant ultrasound spectroscopy. *IEEE Transactions on Ultrasonics, Ferroelectrics and Frequency Control* 55 (2): 476–488.

45. Cao, H., Schmidt, V. H., Zhang, R., Cao, W., and Luo, H. 2004. Elastic, piezoelectric, and dielectric properties of $0.58Pb(Mg_{1/3}Nb_{2/3})O_3$-$0.42PbTiO_3$ single crystal. *Journal of Applied Physics* 96 (1): 549–554.

46. Erturk, A., Bilgen, O., and Inman, D. J. 2008. Power generation and shunt damping performance of a single crystal lead magnesium niobate-lead zirconate titanate unimorph: Analysis and experiment. *Applied Physics Letters* 93(22): 224102.

47. Kobor, D. 2005. Synthèse, dopage et caractérisation de monocristaux ferroélectriques type PZN-PT par la méthode du flux. PhD thesis, Doc'INSA-INSA de Lyon.

48. Zhang, R., Jiang, B., Jiang, W., and Cao, W. 2006. Complete set of elastic, dielectric, and piezoelectric coefficients of $0.93Pb(Zn_{1/3}Nb_{2/3})O_3$-$0.07PbTiO_3$ single crystal poled along 011. *Applied Physics Letters* 89:242908.

49. Jiang, W., Zhang, R., Jiang, B., and Cao, W. 2003. Characterization of piezoelectric materials with large piezoelectric and electromechanical coupling coefficients. *Ultrasonics* 41 (2): 55–63.

50. Lefeuvre, E., Sebald, G., Guyomar, D., Lallart, M., and Richard, C. 2009. Materials, structures and power interfaces for efficient piezoelectric energy harvesting. *Journal of Electroceramics* 22 (1): 171–179.

51. Muralt, P., Antifakos, J., Cantoni, M., Lanz, R., and Martin, F. 2005. Is there a better material for thin film BAW applications than AlN? In *Ultrasonics Symposium, 2005 IEEE* 1:315–320.

52. Chen, Q., Quin, L., and Wang, Q. M. 2007. Property characterization of AlN thin films in composite resonator structure. *Journal of Applied Physics* 101 (8): 084103.

53. Marzencki, M. March 2007. Conception de microgénrateurs intégrés pour systèmes sur puce autonomes. PhD thesis, Université Joseph Fourier Grenoble I, Laboratoire TIMA, Grenoble, France.

54. Carlotti, G., Socino, G., Petri, A., and Verona, E. 1987. Elastic constants of sputtered ZnO films. In *IEEE 1987 Ultrasonics Symposium*, 295–300.

55. Kanno, I., Kotera, H., and Wasa, K. 2003. Measurement of transverse piezoelectric properties of PZT thin films. *Sensors and Actuators A: Physical* 107 (1): 68–74.
56. Muralt, P. 1997. Piezoelectric thin films for MEMS. *Integrated Ferroelectrics* 17 (1–4): 297–307.
57. Wasa, K., Matsushima, T., Adachi, H., Kanno, I., and Kotera, H. 2012. Thin-film piezoelectric materials for a better energy harvesting MEMS. *Journal of Microelectromechanical Systems* (99):1–7.
58. Wasa, K., Ito, S., Nakamura, K., Matsunaga, T., Kanno, I., Suzuki, T., Okino, H., Yamamoto, T., Seo, S. H., and Noh, D. Y. 2006. Electromechanical coupling factors of single-domain $0.67Pb(Mg_{1/3}Nb_{2/3})O_3$-$0.33PbTiO_3$ single-crystal thin films. *Applied Physics Letters* 88:122903.
59. Baek, S. H. et al. 2011. Giant piezoelectricity on Si for hyperactive MEMS. *Science* 334 (6058): 958–961.
60. Guyomar, D., and Lallart, M. 2011. Recent progress in piezoelectric conversion and energy harvesting using nonlinear electronic interfaces and issues in small scale implementation. *Micromachines* 2 (2): 274–294.
61. Lallart, M., and Guyomar, D. 2011. Nonlinear energy harvesting. *IOP Conference Series: Materials Science and Engineering* 18 (9): 092006.
62. Shu, Y. C., Lien, I. C., and Wu, W. J. 2007. An improved analysis of the SSHI interface in piezoelectric energy harvesting. *Smart Materials and Structures* 16:2253.
63. Lien, I. C., Shu, Y. C., Wu, W. J., Shiu, S. M., and Lin, H. C. 2010. Revisit of series-SSHI with comparisons to other interfacing circuits in piezoelectric energy harvesting. *Smart Materials and Structures* 19 (12): 125009.
64. Beeby, S. P., Torah, R. N., Tudor, M. J., Glynne-Jones, P., O'Donnell, T., Saha, C. R., and Roy, S. 2007. A micro electromagnetic generator for vibration energy harvesting. *Journal of Micromechanics and Microengineering* 17 (7): 1257.
65. Muralt, P., Marzencki, M., Belgacem, B., Calame, F., and Basrour, S. 2009. Vibration energy harvesting with PZT micro device. *Procedia Chemistry* 1 (1): 1191–1194.
66. Morimoto, K., Kanno, I., Wasa, K., and Kotera, H. 2010. High-efficiency piezoelectric energy harvesters of C-axis-oriented epitaxial PZT films transferred onto stainless steel cantilevers. *Sensors and Actuators A: Physical* 163 (1): 428–432.
67. Renaud, M., Elfrink, R., Op het Veld, B., and van Schaijk, R. 2010. Power optimization using resonance and antiresonance of MEMS piezoelectric vibration energy harvesters. In *Proceedings of the 10th Workshop on Micro and Nanotechnology for Power Generation and Energy Conversion Applications—PowerMEMS 2010* December 1–3, Leuven, Belgium, technical digest oral sessions, 23–26. Micropower Generation Group, imec/Holst Centre.
68. Halvorsen, E. 2013. Fundamental issues in nonlinear wideband-vibration energy harvesting. *Physical Review E* 87:042129.
69. Kasyap, A. 2007. Development of MEMS-based piezoelectric cantilever arrays for vibrational energy harvesting. PhD thesis, University of Florida, Department of Aerospace Engineering, Mechanics and Engineering Science, Gainesville.
70. Hu, Y., Xue, H., and Hu, H. 2007. A piezoelectric power harvester with adjustable frequency through axial preloads. *Smart Materials and Structures* 16:1961.
71. Cammarano, A., Burrow, S. G., Barton, D. A. W., Carrella, A., and Clare, L. R. 2010. Tuning a resonant energy harvester using a generalized electrical load. *Smart Materials and Structures* 19:055003.
72. Marinkovic, B., and Koser, H. 2009. Smart sand—A wide bandwidth vibration energy harvesting platform. *Applied Physics Letters* 94 (10): 103505.
73. Marzencki, M., Defosseux, M., and Basrour, S. 2009. MEMS vibration energy harvesting devices with passive resonance frequency adaptation capability. *Journal of Microelectromechanical Systems* 18 (6): 1444–1453.

74. Sebald, G., Kuwano, H., Guyomar, D., and Ducharne, B. 2011. Experimental Duffing oscillator for broadband piezoelectric energy harvesting. *Smart Materials and Structures* 20:102001.

75. Sebald, G., Kuwano, H., Guyomar, D., and Ducharne, B. 2011. Simulation of a duffing oscillator for broadband piezoelectric energy harvesting. *Smart Materials and Structures* 20:075022.

76. Hajati, A., Bathurst, S. P., Lee, H. J., and Kim, S. G. 2011. Design and fabrication of a nonlinear resonator for ultra wide-bandwidth energy harvesting applications. In *2011 IEEE 24th International Conference on Micro Electro Mechanical Systems (MEMS)*, 1301–1304.

77. Hajati, A. 2011. Ultra wide-bandwidth micro energy harvester. PhD thesis, Massachusetts Institute of Technology, Department of Electrical Engineering and Computer Science, Boston.

78. Erturk, A., Hoffmann, J., and Inman, D. J. 2009. A piezomagnetoelastic structure for broadband vibration energy harvesting. *Applied Physics Letters* 94:254102.

79. Lin, J. T., Lee, B., and Alphenaar, B. 2010. The magnetic coupling of a piezoelectric cantilever for enhanced energy harvesting efficiency. *Smart Materials and Structures* 19:045012.

80. Andò, B., Baglio, S., Trigona, C., Dumas, N., Latorre, L., and Nouet, P. 2010. Nonlinear mechanism in MEMS devices for energy harvesting applications. *Journal of Micromechanics and Microengineering* 20:125020.

81. Blystad, L. C. J., and Halvorsen, E. 2011. An energy harvester driven by colored noise. *Smart Materials and Structures* 20:025011.

82. Blystad, L. C. J., and Halvorsen, E. 2011. A piezoelectric energy harvester with a mechanical end stop on one side. *Microsystem Technologies* 17 (4): 505–511.

83. Blystad, L. C. J., Halvorsen, E., and Husa, S. 2010. Piezoelectric MEMS energy harvesting systems driven by harmonic and random vibrations. *IEEE Transactions on Ultrasonics, Ferroelectrics and Frequency Control* 57 (4): 908–919.

84. Liu, H., Tay, C. J., Quan, C., Kobayashi, T., and Lee, C. 2011. Piezoelectric MEMS energy harvester for low-frequency vibrations with wideband operation range and steadily increased output power. *Journal of Microelectromechanical Systems* 20 (5): 1131–1142.

85. Soliman, M. S. M., Abdel-Rahman, E. M., El-Saadany, E. F., and Mansour, R. R. 2008. A wideband vibration-based energy harvester. *Journal of Micromechanics and Microengineering* 18:115021.

86. Badel, A. 2005. Récupération d'énergie et contrôle vibratoire par éléments piézoélectriques suivant une approche non linéaire. PhD thesis, Université de Savoie, Chambéry, France.

Index

A